Michael Faraday and the Electrical Century

Iwan Rhys Morus

REVOLUTIONS IN SCIENCE
Published by Icon Books UK

Published in the UK in 2004 by Icon Books Ltd,
Grange Road, Duxford, Cambridge CB2 4QF
e-mail: info@iconbooks.co.uk
www.iconbooks.co.uk

Sold in the UK, Europe, South Africa and Asia
by Faber and Faber Ltd, 3 Queen Square,
London WC1N 3AU
or their agents

Distributed in the UK, Europe, South Africa and Asia
by TBS Ltd, Frating Distribution Centre, Colchester Road,
Frating Green, Colchester CO7 7DW

Published in Australia in 2004
by Allen & Unwin Pty Ltd,
PO Box 8500, 83 Alexander Street,
Crows Nest, NSW 2065

Distributed in Canada by
Penguin Books Canada,
10 Alcorn Avenue, Suite 300,
Toronto, Ontario M4V 3B2

ISBN 1 84046 540 9

Series editor: Jon Agar

Originating editor: Simon Flynn

Typesetting by Hands Fotoset

Printed and bound in the UK by
Mackays of Chatham

Dedication

This book was written for my darling wife Bridgheen, who died before the final chapter was finished. I would like to dedicate it to her memory.

Contents

List of Illustrations

PROLOGUE

So who was Michael Faraday? Faraday, many would say, was the 19th century's most famous scientist. He is certainly one of the period's most familiar scientific names today. After all, he is one of the few British scientists to have appeared on a banknote. We celebrate Faraday for his role in bringing about a scientific and technological revolution that helped usher in the modern world. Faraday's experiments transformed the science of electricity. He invented the electric motor. He investigated the relationship between electricity and magnetism, showing that currents of electricity could be produced by a moving magnet – an insight that lay at the heart of the 19th-century electrical power industry. Without it, things we take completely for granted, like electric lights that go on and off at the mere flick of a switch, would simply be impossible. We live now in a world in which we are surrounded by things electrical. It is hard to imagine a world in which such things did not exist, or where their very appearance seemed magical. That was the early Victorian world in which Faraday lived, however. One slightly over-excited commentator at the time (the Bishop of Llandaff, no less) thought electricity 'far exceeds even the feats of pretended magic and the wildest fictions of the East'. It would achieve 'a thousand times more than what all the preternatural powers which men have dreamt of and wished to obtain were ever imagined capable of doing'.

At the beginning of the 19th century there was no electrical industry. By its end, the rise of electricity seemed inexorable. During that century, electricity transformed long-distance communication with the invention of the telegraph. By the end of the century, power generation and transport were undergoing their own electrical revolutions too. Faraday's Victorian contemporaries were fascinated by electricity. To a degree that may seem incomprehensible to their cynical 21st-century descendants, they felt that electricity symbolised their century's progressive optimism. But was electricity just down to Faraday? Faraday was certainly celebrated by his contemporaries because of his contributions to that most progressive of sciences – and the scientific links between Faraday's discoveries and the new technologies on which the growing Victorian electrical industry depended may be clear. But there was a little more to it than that, as many Victorians recognised. Just what the relationship was between science and industry was a real hot potato for Faraday's contemporaries. They argued endlessly over whether science and industry were part of the same package or whether they should remain forever separate. Faraday himself, it should be noted, would have been insulted at the suggestion that his main claim to fame was that he had invented the electric motor. In his view, he had done no such thing. So how responsible was Faraday, really, for inventing the electrical century? That is what this book sets out to discover.

In much the same way that Jeeves was Bertie Wooster's 'gentleman's gentleman', Faraday was Victorian gentlemanly society's scientist. Not a gentleman himself – as we shall see – his science was at the service of genteel society. Indeed, in large degree he defined what polite Victorian culture thought science was all about. One of the things that defined that kind of polite science was that it was about discovery rather than invention. Natural philosophers like Faraday discovered things. Other, lesser beings (and ones with rather dirtier hands) invented. This is not to suggest that Faraday did not think science should be useful (he did) or that he did not think invention was a worthwhile activity (again, he did). The point is that invention was a very different kind of activity from discovery, and one that was, in Faraday's view at least, less dignified. Others obviously disagreed, and maintained very different views about the relationship between the science of electricity and its technology. As far as someone like Faraday's contemporary William Sturgeon was concerned, for example, discovery and invention were exactly the same thing, and being an inventor just as good a claim to scientific fame as being a discoverer. We need to look at the Sturgeons as well as the Faradays if we really want to make sense of the electrical century.

One aspect of Faraday's career was certainly as celebrated during his lifetime as it is now. Faraday was as famous then as he is now for being a self-made man. This is why he was Margaret Thatcher's favourite scientist, after all. He embodied what she imagined to be the cardinal Victorian values of self-discipline and self-help. Faraday had been successful at doing without society, getting to the top despite the lack of a formal scientific education and a humble background. Ironically, for much the same reason, Faraday has also been held up as a working-class hero, battling against the odds and a conservative scientific establishment to emerge triumphant despite the difficulties stacked against him. All of Faraday's 19th-century biographers certainly made great play of this aspect of his career. It did indeed fit in well with some Victorian ideas about the importance of self-help. One Victorian socialite saw Faraday as an example of how the right kind of humility before nature could transcend Victorian social barriers. 'Sixteen quarterings of pure Norman ancestry', enthused the distinctly blue-blooded Cornelia Crosse, 'could not have made Michael Faraday, the blacksmith's son, a finer gentleman than he was by nature'. As we shall see, however, this public image as one of nature's gentlemen did not come about purely serendipitously, or simply through hard work in the laboratory. It was an image that Faraday carefully cultivated.

To understand Faraday, his contemporary success and even his modern-day popularity, we need to look behind the more familiar pictures. We need to see how they were put together, and why. So this is not, therefore, a conventional biography. It certainly does not aim to give the kind of detailed and exhaustive account of Faraday's life that a proper biography should. This book will look instead at particular episodes in Faraday's career in the context of his times. There will even be chapters in which Faraday features barely, or not at all. The years in which Faraday grew up on the streets of London were turbulent ones. The country was at war. The dangerous spectre of revolutionary France loomed over the Channel. New scientific – as well as political – ideas had their roots in the conflict. There were often close links between the two. Conservative critics frequently lambasted the new science of electricity, in which Faraday would eventually make a name for himself, for its radical and French connotations. In fact, one of the many things that Faraday achieved was to make electricity respectable and English. Looking at some of this context will help give us a sense of where Faraday and his science came from – of what it meant to him and to his contemporaries. It will also remind us that Faraday did not work in a vacuum. There were other men of science whose views of electricity and how it should be practised differed drastically from his.

Science itself was in a state of flux during the first half of the 19th century. There was no tried and tested path to becoming a man of science – the term 'scientist' itself was not even coined until 1833. There were no university degrees, no PhD programmes, no postdoctoral fellowships. There were certainly very few positions where a man of science could expect to be paid for his services. Men of science were, by and large, leisured gentlemen – those with the time and money to dabble in natural philosophy. New institutions were, however, being established, and ambitious young Turks were angling to take over and reform the older ones like the prestigious Royal Society. People argued over whether science should be a vocation or a career. They debated whether science should be economically useful or if it should be recognised as a good thing in itself. Should membership of prestigious scientific bodies be decided on merit or on social standing? This was the world through which Faraday – who was certainly from the wrong side of the tracks as far as most of his contemporaries in science were concerned – had to negotiate a path for himself. Men of science and engineers ended up transforming 19th-century society. Electricity in particular was at the very heart of this brave new modern world. There was nothing inevitable about any of this. Men like Faraday had to work hard to carve out a niche for themselves in Victorian society.

Looking at Faraday's career can therefore help us better understand the electrical century in which he lived and worked as well. Understanding how Faraday went about forging a supremely successful scientific career from what were, by any standards, highly unpromising beginnings can help us understand the complex and hierarchical Hanoverian and Victorian scientific society through which he moved. By following Faraday around, we can try to uncover the networks of influence and patronage that made 19th-century science possible. We will see what kinds of resources – both material and social – were available to a budding young philosopher from the wrong side of the tracks. Genius (a Romantic idea that became popular at just about this time, and of which Faraday clearly had plenty) was never going to be enough to guarantee a successful scientific career in a society in which being a gentleman was an important and typical characteristic of the man of science. By understanding what Faraday saw himself as trying to achieve, we will be able to get a better sense of what his contributions – and those of a number of others – to the electrical century really were. We will see how Faraday forged his own image as humble investigator of nature, and how others remade him in their own image and for their own reasons.

Part I: Growing Up in Scientific London

· Chapter 1 ·

The Streets of London

Faraday was born on 22 September 1791 in Newington, south of the Thames in Surrey, near the Elephant and Castle and within shouting distance of the Old Kent Road. His father James Faraday was a journeyman blacksmith, recently married to Margaret Hastwell, a farmer's daughter. They were recent arrivals in London, being originally from the village of Clapham in Yorkshire. They did not stay long in Newington. By the time Faraday was five they were settled into rooms above a coach house in Jacob's Well Mews, off Manchester Square. James worked in a smithy in nearby Welbeck Street. Faraday's parents were followers of the Sandemanians, a small non-conformist sect outside the dominant Church of England, and his father formally joined the Sandemanian Church shortly after their arrival in London. Faraday would be a devout Sandemanian throughout his life, becoming an elder of the Church but also at one period being temporarily expelled from its ranks. In 1809, with Faraday in his late teens, the family moved again, to Weymouth Street near Portland Place, where James Faraday, never in good health, died about a year later. Life for the Faradays was a struggle. During one period of high corn prices in 1801 – as the war with France demanded its economic pound of flesh – young Michael was forced to subsist on a single loaf of bread a week. James Faraday's ill health meant that he was often unable to complete a day's work.

Faraday's early education was basic. As he recalled himself: 'My education was of the most ordinary description, consisting of little more than the rudiments of reading, writing, and arithmetic at a common day-school.' He was probably quite lucky to have received even as much of a rudimentary education as that. Times were hard and even the most basic education cost money. Literacy rates amongst London's poorer classes at the beginning of the 19th century were not high. As a journeyman blacksmith – someone who worked for a master rather than being an independent master craftsman himself – it is unlikely, particularly given his recurrent ill health, that James Faraday earned much more than twenty shillings a week at the best of times. Little more is known about Faraday's early childhood. When he was not at school, his time was passed at home or on the streets, playing marbles in nearby Spanish Place or looking after his little sister playing in Manchester Square. Given his family's relative poverty, it is unsurprising that the young Michael Faraday was expected to make his own economic contribution to the family finances as soon as he was able. In 1804, when Faraday was thirteen, a bookseller, George Riebau, who kept a shop just around the corner from Jacob's Well Mews at No 2 Blandford Street, hired him as an errand boy.

On 7 October 1805, after he had been working as his errand boy for about a year, Faraday was indentured as an apprentice to George Riebau to learn the trade of bookbinder and stationer. Faraday

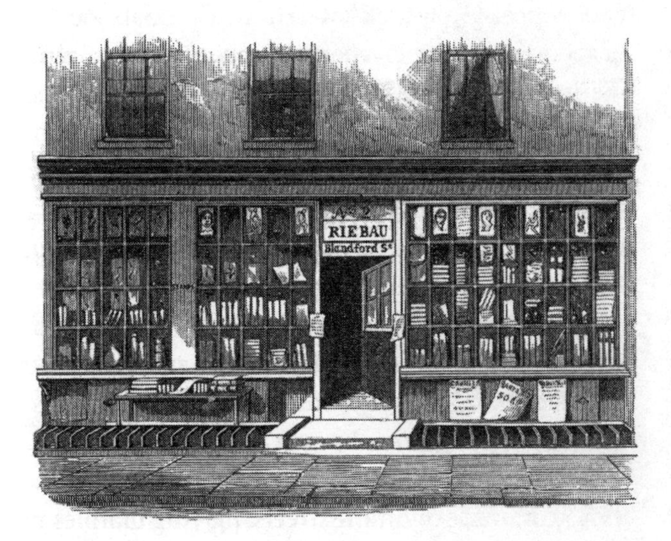

Illustration 1: George Riebau's bookshop on Blandford Street. Faraday worked here for several years as a bookbinder's apprentice, and first started to become interested in electricity.

and his family were lucky. Apprenticeships into a good trade like this were highly prized and often kept within the family circle. They were even luckier that Riebau agreed to waive any payment for the apprenticeship, in view of Faraday's 'faithful service' as an errand boy. As an apprentice, Faraday was expected to live with his master and mistress for a number of years and learn the rudiments of his master's trade. At the end of the period – typically seven years – he would produce a masterpiece demonstrating his proficiency in the craft, and would become a journeyman, judged competent to seek employment in the trade in his own right. Riebau was clearly a good master and Faraday seems to have flourished under his tutelage. About half way through the apprenticeship, his father noted that his son was 'very active at learning his business' after a few teething troubles at the beginning of his term. He had 'rather got the head above water' and had two junior apprentices under his command.

London booksellers' shops were important and interesting places to be in the turbulent early years of the 19th century. They were far more than simply places where Londoners went to buy books. They were also significant meeting points for the metropolis's underground network of radical political agitators and rabble-rousers. Early 19th-century booksellers, like their 18th- and 17th-century predecessors, did more than just sell books. They often published them as well. There was a flourishing black market trade in the early 19th century in politically seditious and pornographic pamphlets and tracts. Political hacks and subversives, stridently demanding the rights of man, published filthy broadsides, condemning the corruption and mutual back-scratching of Regency culture in general and of William Pitt the Younger's reactionary and repressive regime in particular. George Riebau certainly had a hand in this under-the-counter trade. He was part of a network of illicit publishers and pamphleteers churning out a steady stream of anti-establishment literature aimed at radicalising the working classes and chipping away at the bastions of power and privilege. Working away in his shop, the teenage Faraday would have witnessed London's radical underworld in action. He would have seen the way that some of its members turned to science as a powerful weapon in the fight for social justice and political emancipation. In later life, Faraday was to have none of it.

What he also found out about – and clearly fell in love with – was science. As Faraday himself recalled: 'Whilst an apprentice I loved to read the scientific books which were under my hands, and, amongst them, delighted in Marcet's "Conversations in Chemistry", and the electrical treatises in the "Encyclopaedia Britannica".' Jane Marcet's *Conversations in Chemistry* (first published in 1806) was written in dialogue form, following a conversation on chemistry between a governess and her two female charges. It introduced the bookbinder's apprentice to the chemical philosophy of Humphry Davy, who was to play a vital part in his life a few years later. Early 19th-century books were usually bought unbound. It was Faraday's job to bind them. He clearly took advantage of the opportunity to read them as well, being an enthusiastic and indiscriminate reader. 'Do not suppose that I was a very deep thinker', he warned his Royal Institution successor, John Tyndall, 'or was marked as a precocious person. I was a very lively imaginative person, and could believe in the "Arabian Nights" as easily as in the "Encyclopaedia". But facts were important to me, and saved me. I could trust a fact, and always cross-examined an assertion.' Inspired by his readings and what he found out about the latest developments in electricity (a topic that would certainly have fascinated the political radicals dropping in to Riebau's shop as well), he started experimenting. 'I made such simple experiments in chemistry as could be defrayed in their expense by a few pence per week, and also constructed an electrical machine, first with a glass phial, and afterwards with a real cylinder, as well as other electrical apparatus of a corresponding kind.' With his master's permission and money donated by his older brother, Robert, he also started attending his first lectures.

· CHAPTER 2 ·

SCIENTIFIC LONDON

So what was out there for an impecunious young apprentice looking to discover science in early 19th-century London? Regency London had a flourishing scientific culture. A whole range of scientific institutions of various kinds prospered. There was a thriving culture of popular scientific lecturing. A scientific tourist with enough time on his hands, plenty of money in his purse and – crucially – the right introductions in his coat pocket could look forward to rubbing shoulders with some of the greatest scientific minds in Europe and seeing some real scientific wonders. A steady diet of popular scientific lectures and spectacular demonstrations had been a staple part of the metropolis's coffee-house culture since at least the early 18th century. Such events continued to be a draw for fashionable and genteel audiences in the opening decades of the 19th century as well. Not only home-grown scientific lecturers, but a steady stream of foreign performers turned up on London's scientific stages. Our scientific tourist, if he had visited London in 1803, could have witnessed the

Italian natural philosopher Giovanni Aldini at work, for example. The nephew of Luigi Galvani (the discoverer of animal electricity – the eponymous galvanism) carried out electrical experiments on the body of an executed murderer, presided over by the President of the Royal College of Surgeons.

London science had its own social circle and its own hierarchy of institutions and activities. At its pinnacle at the beginning of the 19th century was the Royal Society. Founded in 1662 with a royal charter from Charles II, the Royal Society was from its beginnings a society of scientific gentlemen devoted to the increase of natural knowledge. In 1810 its President, Sir Joseph Banks, had been at the helm for more than 30 years. He had made a name for himself as a botanist and natural historian during Captain James Cook's expeditions to the South Seas. Independently wealthy, he had used his money, his connections and the name he had made for himself as Cook's botanist to place himself at the centre of a powerful network of scientific patronage with the Royal Society at its core. To Banks's enemies, of whom there were quite a few by 1810, the Royal Society under his nepotistic regime typified everything that was wrong with metropolitan (and English) science. It was corrupt – Banks held most of the meagre public purse strings of English science in his own hands and doled the money out to his own protégés. It was aristocratic and dilettante in its attitudes and interests. For a new generation of meritocratically minded gentlemen of science who were starting to come to the fore of metropolitan science during the 1810s, the Royal Society looked like a fruit ripe for the picking.

The Royal Society's Fellows – those who could put the magic letters FRS after their names – were a self-proclaimed philosophical élite. During the early 19th century the Society held its meetings at Somerset House on the Strand, where they had moved from a house in Crane Court in 1780. Meetings were ceremonial affairs, usually presided over by Sir Joseph Banks himself. Papers were read out by one of the Society's secretaries rather than by their authors, before being formally presented to the Society. Elections to the prestigious Fellowships also took place at meetings, with potential Fellows being nominated by existing ones and put up to a vote of those present. There was no particular requirement that a potential Fellow should be an active man of science. As far as Banks and his cronies were concerned, power and patronage were just as good reasons as a scientific reputation for being made an FRS. The Royal Society held its meetings weekly during the winter months. There were no meetings during the summer when many of the Fellows would be found at their country estates rather than in the city. A bookbinder's apprentice like Michael Faraday would have stood little or no chance of ever being admitted into the Royal Society's Somerset House chambers. It was strictly a society for scientific gentlemen.

Only slightly lower down the social scale of London science during this period would have been the Royal Institution. The RI was a recent foundation, having been established only a few years previously in 1799 by a coterie of gentlemen, with the Royal Society President, Sir Joseph Banks, foremost in their midst. Inspiration for the new scientific institution had been provided by the American émigré and loyalist, Benjamin Thompson, who had been obliged to flee the former Colonies with the advent of American independence. Thompson's plan had been to establish an institution for diffusing knowledge and new inventions and improvements, as well as 'for teaching, by means of philosophical lectures and experiments, the application of science to the common purposes of life'. A mansion in fashionable Albemarle Street was purchased to provide a home for the new institution. The original plan had been for an institution that would provide scientific lectures aimed at all social classes. In his early drawings, Thomas Webster, the Institution's architect, incorporated a staircase leading directly from the outside to the lecture theatre so that the city's rude mechanicals could gain entry without mixing with their betters in the lobby. The continuing war with

France and radical unrest at home soon started to make such plans for workers' education look decidedly suspect, however. Webster found himself being 'asked rudely what I meant by instructing the lower classes in science ... it was resolved upon that the plan must be dropped as quietly as possible. It was thought to have a political tendency.'

The Royal Institution spawned a number of competitors, though none of them ever really succeeded in emulating the social and scientific exclusivity of the fashionable institution on Albemarle Street. The London Institution was founded in 1805 by a splinter-group from the Royal Institution, unhappy with that body's aristocratic ethos and emphasis on agricultural improvement. As hard-headed city businessmen and professionals, what they wanted to foster was the union of science and commerce. Science and commerce combined, Charles Butler insisted in his oration at the laying of the foundation-stone of the Institution's Finsbury Circus building, would 'record the heavens, delve the depths of the earth, and fill every climate that encourages them with industry, energy, wealth, honour, and happiness:- To civilization, to virtue, to religion, they open every climate; they land them on every shore; they spread them on every territory.' As well as the London Institution, other competitors to the Royal included the Surrey Institution near Blackfriars Bridge, where the chemist Frederick Accum lectured (and which possessed 'an excellent library and a still more decent Laboratory'), and the Russell Institution in Bloomsbury, both founded in 1808. As well as drawing their inspiration from the Royal Institution, these smaller scientific institutions also modelled themselves on the growing body of provincial literary and philosophical societies that were so much in vogue during the end of the 18th and the beginning of the 19th centuries. The first of these, the Manchester Literary and Philosophical Society, had been founded in 1781.

Faraday's chances of getting inside the Russell or Surrey Institutions, or even the London Institution, while probably higher than his hopes of entering the Royal Institution's gilded portals, were still not very promising. By the 1810s, London boasted its own Literary and Philosophicals as well. The Hackney Literary and Philosophical Society, founded in 1810, met every Tuesday for scientific conversation and lectures on subjects ranging from literature through chemistry and galvanism to mathematics. The Marylebone Lit. and Phil. was founded a little later, with a lecture theatre in Portman Square that could accommodate 600 people. In 1806, the London Chemical Society was established at Old Compton Street in Soho with 60 members on the books. It had its own well-equipped chemical laboratory and provided a good institutional base for the likes of Frederick Accum before he moved on to the Surrey Institution. The London Philosophical Society had its origins as early as 1794 in a group of enthusiasts centred around the watch- and instrument-maker Samuel Varley. By 1811 the Philosophical Society of London occupied the Royal Society's old quarters in Crane Court. Unlike their Royal predecessors, the new incumbents promised 'to foster genius, to eradicate unphilosophic prejudice, to increase the knowledge of nature, and most, of man'.

Down at the very lower end of the scientific market-place – and therefore probably most accessible to someone in Faraday's position – scientific shows of various kinds proliferated. The popular lecturer Adam Walker performed regular courses of twelve lectures each on 'Natural and Experimental Philosophy' at such unlikely venues as the King's Head in the City and the Theatre Royal in Haymarket. Entry to his lectures cost a guinea for the course: still a considerable sum for an apprentice boy. Walker founded a mini-dynasty of popular scientific lecturers who were still active more than a decade later. Another prolific and popular performer was Mr Hardie, a surgeon based at Great Portland Square. Hardie gave his performances on anything from astronomy, chemistry or electricity to mathematics and medicine, at his own Theatre of Science on Pall Mall. As well as extended lecture courses, he offered single performances tailored to the impecunious enthusiast. Lecturers such as these, or others who performed at venues like the Lyceum or the Theatre of Astronomy on Catherine Street, were unabashedly popular, adapting their presentations to the expectations of their audiences and taking advantage of every opportunity to attract new customers. They were theatrical performances in every sense of the word.

This range of scientific activities and institutions sustained a burgeoning group of more or less peripatetic scientific lecturers in London. Very few popular scientific lecturers confined their activities to a single institution. They moved around from place to place and offered courses of their own independently as well. Success in London could provide the basis for a lucrative tour of the provinces. Similarly, lecturers who had made a reputation for themselves in the provinces gravitated towards London and its possibilities. A good example is Thomas Garnett, who had lectured across the Midlands and at Glasgow's Anderson's Institution before coming to London, initially as a Professor at the Royal Institution. After falling out with Benjamin Thompson he set himself up as an independent, working from his base on Great Marlborough Street and at Tomm's Coffee House in Cornhill. His protégé, George Birkbeck (later to play a key role in establishing the London Mechanics'

Illustration 2: William Walker lecturing on astronomy at the English Opera House. Walker, the product of a mini-dynasty of spectacular scientific lecturers, was one of Faraday's favourite performers.

Institution), similarly moved from the Anderson's Institution to try his luck on the London lecturing circuit. Many players on the metropolitan popular scientific scene, like Garnett and Birkbeck (and the prolific Mr Hardie), were medical men and looked to providing lectures to medical students for a large part of their income.

Chemistry was another popular subject, again at least partly because of its attraction to medical students, but also because of its commercial potential. William Thomas Brande, a friend of Frederick Accum of the Surrey Institution, lectured on chemistry at the Great Windmill Street Theatre of Anatomy, paying particular attention to the application of chemistry to 'the Arts and Manufactures'. Brande was one of relatively few popular lecturers making a living towards the lower end of London's scientific market-place who succeeded in making it to the higher echelons. Brande eventually became an FRS, as well as Professor of Chemistry at the Royal Institution. London's popular lecturers – the successful ones at least – kept a careful eye on the needs and expectations of their audiences. This is why medical subjects and chemistry (and geology and mineralogy too) were such common topics. They fulfilled the popular demand for useful knowledge. Not all popular lectures had to be so straightforwardly utilitarian, however. London's lecture audiences liked a good show, and sciences such as astronomy or electricity could draw good crowds for that reason. Electricity in particular, with the invention of new devices such as Alessandro Volta's voltaic pile in 1800, was a reliable crowd-pleaser. It was an excellent source of spectacular displays of nature's powers – and of the lecturer's own powers over nature.

· Chapter 3 ·

First Steps in Science

So what do we know about Michael Faraday's early forays into London scientific culture? We know that he took full advantage of his position in George Riebau's bookshop to read scientific books and journals as they came his way. We know that he constructed various pieces of scientific equipment – such as an electrical machine for generating static electricity – and carried out experiments. This in itself was a significant achievement for a working-class youth with little in the way of an education. Michael Faraday had already acquired the determined drive for self-improvement through science that was to prove so crucial to his later career. Riebau was sympathetic to his apprentice's interests and allowed him to use the shop's facilities. As one of Faraday's friends at the time, Benjamin Abbott, recalled: 'In the shop where M.F. worked as an apprentice was a small fire place used for warming the room & heating the tools used in gilding. By the aid of some milled boards he contrived, when he wished it, to convert this into a furnace & thus to make experiments in Metallurgy; whilst on the

Mantelpiece were placed sundry little voltaic piles & other matters constituting as it were a miniature Laboratory. His employer took some interest in these pursuits & during the latter years of his apprenticeship allowed him the privilege of banishing a junior from the shop after the hours of work, & thus securing himself from interrogation.'

To what extent Faraday availed himself of London's popular lectures we do not know. It seems unlikely, given his interests and commitment to science, that he attended none of them. It is known that he attended and enjoyed William Walker's astronomical lectures at the English Opera House, describing them as amongst the best in London, but this was probably at a later date. Lack of money would have been a major problem, nevertheless, even at the cheaper end of the spectrum of offerings. Thanks to his brother Robert, however, he got his chance in 1810 to attend a course of popular lectures offered by the scientific lecturer and silversmith John Tatum at his house, 53 Dorset Street, near Fleet Street. Tatum was a customer of George Riebau's and it was through this association (as well as his brother's shillings) that Faraday got his opportunity to attend. Tatum's lectures were largely aimed at young men in Faraday's position – avid for scientific knowledge and hungry for self-improvement. They cost a shilling per lecture. This was far from being a negligible sum, and constituted a significant financial commitment on Faraday's (or, rather, on his brother Robert's) part. Tatum's lectures would have provided Faraday with a grounding in the basics of natural philosophy as well as a survey of the latest discoveries. Just as importantly, they also introduced him to a circle of like-minded friends and fellow-aspirants.

As well as giving scientific lectures to young hopefuls like Faraday, Tatum was also the guiding force behind the City Philosophical Society. Founded in 1808, this society was in large part a continuation of Tatum's lectures. It was intended to provide a forum that would enable the regulars at those lectures to further increase their scientific knowledge and engage in mutual self-improvement. The Society met every Wednesday evening at Tatum's house. Meetings alternated between hearing one of the members give a lecture on a scientific subject and discussion meetings where the latest scientific issues were aired between members and their guests. Significantly for such contentious times, political and religious discussion was explicitly forbidden by the Society's constitution. This provision did nothing to help the Society a few years later when it found itself on the wrong end of the law following the passing of the Seditious Meetings Act in 1817. Questions had to be asked in the House before the City Philosophical Society eventually received its licence to meet. In the meantime, however, Faraday seems to have been a regular presence at City Philosophical Society meetings from about 1810 or so onwards. By 1812 he was making an active effort to become a member, which would have allowed him to put the letters MCPS after his name. We do not know whether his efforts then were successful, but he certainly became a member a few years later.

Faraday's determination to improve himself through science is evident in the correspondence he soon entered upon with Benjamin Abbott, a fellow regular at the City Philosophical Society with whom Faraday had struck up a particular friendship, founded, as Abbott expressed it, 'on the mutual esteem of each for the other, & on a good degree of similarity in our youthful aspirations'. The correspondence began in July of 1812, close to the time when Faraday's apprenticeship came to its end, and he clearly regarded it as yet another element of his drive towards self-improvement. Writing letters regularly to a like-minded friend would help teach him the discipline of expressing himself and his ideas clearly and economically. Faraday used Abbott as a sounding board for his ideas about experimental natural philosophy, scientific lecturing and mental improvement. In 1809, Faraday had read Isaac Watts's *The Improvement of the Mind* and become a convert to Watts's system of methodical self-betterment through mental cultivation. In fact, writing regularly to a like-minded correspondent, as Faraday and Abbott had done, was one of Watts's suggestions for self-improvement. By 1812, Faraday was in fact desperate for self-improvement. His apprenticeship was almost over (it finished on 7 October that year) and a future as a bookbinder seemed thoroughly unattractive. What he wanted was a future in science. He even wrote a begging letter to the aristocratic Sir Joseph Banks at the Royal Society, pleading for his patronage, but received a scribbled 'no answer' in reply.

Faraday's break came about, yet again, through Riebau's bookshop. One of Riebau's regular customers, William Dance, was a member of the Royal Institution. The young apprentice's interests and aspirations were clearly a matter of common knowledge among Riebau's clients, and the upshot was that Dance offered Faraday some tickets to attend Sir Humphry Davy's lectures at the Royal Institution. This was an unprecedented opportunity. Under normal circumstances and without the patronage of a well-wisher like Dance, someone like Faraday would have had little or no chance to watch a scientific superstar such as Humphry Davy in action. Faraday took full advantage of his opportunity. He attended the lectures diligently and, as he had been taught by Tatum and at the City Philosophical Society, as well as in Isaac Watts's self-improving exhortations, he took careful notes. Aged 21, with his apprenticeship over and his appeal to Banks come to nothing, Faraday entered employment as a journeyman bookbinder with Henri de la Roche. A few months later, he was Sir Humphry Davy's laboratory assistant at the Royal Institution. Faraday had it made! He had sent Davy a bound copy of his lecture notes, secured an interview and, having persuaded Davy of his sincerity, landed the job.

PART II:
THE PHILOSOPHER'S APPRENTICE

· CHAPTER 4 ·

A FASHIONABLE PLACE TO BE

To understand why Faraday's strategy worked and just how significant a move it would turn out to be, we need first of all to know a little more about the Royal Institution and, in particular, its flamboyant Professor of Chemistry, Humphry Davy. As we saw in the last chapter, the Royal Institution had been established in fashionable Albemarle Street in 1799, with a brief to bring science to its well-heeled metropolitan clientele. What had started out as a philanthropic endeavour – to try to better the condition of the poor (and, incidentally, stave off revolution) by applying chemistry to agricultural improvement – soon became a purveyor of sanitised scientific knowledge for the upper classes. The Royal Institution had formidable resources at its command. Its building in Albemarle Street contained one of London's first purpose-built scientific lecture theatres. It also contained a laboratory in the basement, taking up the space occupied by the kitchen and outhouses of the original mansion, where its professors could prepare their experiments beforehand. Crucially, the Royal Institution's professors could also draw on the services of six workmen and a mathematical instrument-maker to help them in their experiments. All in all, these were unrivalled resources and made the position of Professor of Chemistry at the Royal Institution a truly enviable one.

The Royal Institution's first Professor of Chemistry was Thomas Garnett, an experienced performer from Glasgow's Anderson's Institution. He soon left, however, following disagreements with the managers over his salary. He was replaced by Humphry Davy, who had recently been hired as his assistant. Born in 1778 in the Cornish town of Penzance, Davy was the son of a down-on-his-luck and impoverished wood-carver. He was apprenticed to a local apothecary and seemed set to make his career as a provincial medical practitioner until family connections brought him into contact with the radical chemist and doctor Thomas Beddoes. Beddoes was in the process of establishing the Pneumatic Institute in Bristol, devoted to curing disease by the judicious inhalation of different kinds of air. The bright and ambitious apothecary's apprentice with an interest in chemistry seemed the ideal assistant, and was duly hired. It was Davy's task at the Pneumatic Institute to experiment with different kinds of air, discovering their effects on the body. Beddoes was a follower of the physician John Brown, who held that most diseases resulted from excessive or deficient excitability and could therefore be treated by an appropriate regime of stimulants and sedatives to restore the proper balance. Davy's experiments on different airs – such as nitrous oxide (better known as laughing gas) – at the Pneumatic Institute were aimed at classifying them as stimulants and sedatives in this way. To some, at least, this was a politically suspect programme. Tory critics reckoned that this sort of chemical and philosophical dabbling was responsible for some of the worst excesses of the French Revolution. They were not entirely wrong, either. Men like Beddoes had every sympathy for the political revolution taking place across the Channel, and certainly saw their chemistry as part of a political campaign to reform society.

Davy's move from the radical Pneumatic Institute, where he associated with Romantic poets like Coleridge and Southey, to the aristocratic Royal Institution was regarded by some of his old friends as a betrayal of principle. It rapidly became clear that the Pneumatic Institute regime of inviting members of the audience to participate in partaking of the various kinds of air during lectures would not work in this new setting. The Royal Institution's genteel audience did not like being made to look silly. Davy soon dropped pneumatic chemistry in favour of a new fluid: galvanism. Galvanism – or voltaic electricity – was very much in vogue during

Illustration 3: A cartoon by James Gillray, lampooning Thomas Garnett's pneumatic performances at the Royal Institution. The young man on the right in the background, wielding a bellows and wearing a manic grin, is Humphry Davy.

the early years of the 19th century. The debates between Luigi Galvani – who contended that electricity could be produced from animal tissues – and his fellow Italian Alessandro Volta – who argued that it was the result of contact between different metals – had resulted in a powerful new tool of experimental inquiry: the voltaic pile (ancestor of the modern electric battery). Davy used the pile to great effect in his lectures, wowing his audience with spectacular shows of shocks and sparks. At the same time, he used the pile to make his reputation as an experimenter, showing how it could be used to discover new elements by ripping the existing ones apart. In so doing, he mounted a strong English challenge to French chemical theorists such as Lavoisier, showing that his list of elements was wrong. He challenged Volta, the pile's inventor, as well, arguing that the pile's electricity was the product of chemical forces rather than the mere contact of metals. This was a style of science that went down very well indeed in parochial, embattled England during the years of the Napoleonic Wars.

Chemistry made Davy's reputation. It also helped make him his fortune. By the time Faraday first encountered him in 1813 he had come a long way from being a down-at-heel Penzance apothecary's apprentice. He had been knighted by George III in 1812. He was the doyen of London science with a European reputation. He had also met his rich widow, Jane Apreece, and married her three days after being knighted. In later life, Faraday would remark rather cattily that he had learned a great deal from Davy – primarily about what not to do. Davy's rise to fame and fortune had certainly made him his fair share of enemies. He was lampooned as a gauche *nouveau riche* who tried to ape his betters instead of knowing his proper place. Davy was certainly keenly aware of his own reputation. Aspiring to be a gentleman, he did everything he could to avoid contact with anything that hinted of trade. He was contemptuous of William Brande, his successor as Professor of Chemistry at the Royal Institution, dismissing him as a 'mercenary' who had 'come from the counter' and 'had no lofty views'. Davy himself had 'come from the counter' as well, of course, but usually did his best to forget it. In transforming himself from apothecary's boy to savant, however, he also transformed the science of electricity from a dubious enterprise tainted by radical connotations into a bulwark of aristocratic science.

Nevertheless, Davy with his humble beginnings was in a far better position than most other men of English science to appreciate the gesture that Faraday made to him in late 1812. As we saw at the end of the last chapter, Faraday, after diligently attending Davy's lectures, carefully bound his notes and sent them to him. What was Faraday doing? Davy, at least, as a former apprentice himself, would have realised at once that this was a deeply symbolic action. Faraday was presenting him with a masterpiece, just as an apprentice presented his master with a masterpiece to signify the end of his apprenticeship and the start of his career as a journeyman. In this case, of course, Faraday was trying to convey a slightly different message. He was using a demonstration of the skills he had learned in one trade as an apprentice bookbinder to petition for entry into another – that of a man of science. In effect he was asking for a second apprenticeship in science. Davy, with his background in provincial artisanal culture, knew exactly what Faraday was asking from him. By taking Faraday as an assistant he would effectively be entering into a contract with rights and obligations on both sides.

· CHAPTER 5 ·

THE GRAND TOUR

Davy was clearly keen to help his young supplicant in any way that he could. He told Faraday that he was 'far from displeased with the proof you have given me of your confidence, and which displays great zeal, power of memory, and attention'. Nevertheless, he had to wait for the right opportunity. Some months after Faraday's appeal, William Payne, one of the Royal Institution's laboratory assistants, was sacked. John Newman, the Institution's philosophical instrument-maker, had accused him of dereliction of duty in failing to attend and assist at William Brande's chemistry lectures. According to Newman at least, a brawl developed and the instrument-maker was injured. Despite his ten years' service, Payne was promptly dismissed and the Royal Institution found itself in need of a new laboratory assistant. Davy immediately recommended Faraday, informing the managers that: 'As far as Sir H. Davy has been able to observe or ascertain, he appears well fitted for the situation. His habits seem good, his disposition active and cheerful, and his manner intelligent.'

Faraday was duly hired as chemical assistant with a wage of 21 shillings a week (shortly thereafter increased to 25 shillings), along with room and board. Faraday had been successful in his quest to become the philosopher's apprentice.

But what did the philosopher's apprentice do? When Davy had asked around amongst his philosophical friends and colleagues for suggestions as to what he should do with his young protégé, one of them, William Pepys, suggested that he should be put to washing bottles. Davy had thought that too degrading, but Faraday's list of official duties was by and large menial: 'To attend and assist the lecturers and professors in preparing for and during lectures. Where any instruments or apparatus may be required, to attend to their careful removal from the model-room and laboratory to the lecture-room, and to clean and replace them after being used, reporting to the managers such accidents as shall require repair, a constant diary being kept by him for that purpose. That in one day in each week he be employed in keeping clean the models in the repository, and that all the instruments in the glass cases be cleaned and dusted at least once within a month.' Not so far from bottle washing after all, maybe. In practice, however, Faraday almost immediately found himself being given the opportunity to help Davy in his own experiments and to prepare chemical compounds. He was being taught the routines and practices of laboratory life by a true expert.

In view of the common perception of Faraday as an entirely self-taught man of science, this last point cannot be over-emphasised. Far from being deprived of scientific training, Faraday was being given one-to-one tuition in the art of experiment by one of the greatest European experimental natural philosophers. Writing his first letter to his friend Benjamin Abbott following his appointment at the Royal Institution, Faraday was full of how he had been 'employed to-day, in part, in extracting sugar from a portion of beetroot, and also in making a compound of sulphur and carbon – a combination which has lately occupied in a considerable degree the attention of chemists'. Being Davy's chemical assistant was not without its dangers. Davy was at the time experimenting with highly volatile compounds of chlorine and nitrogen. In his next letter to Abbott, Faraday described how he had escaped serious injury from 'four different and strong explosions' of such compounds: 'It exploded by the slight heat of a small piece of cement that touched the glass above half an inch above the substance, and on the outside. The explosion was so rapid as to blow my hand open, tear off a part of one nail, and has made my fingers so sore that I cannot yet use them easily. The pieces of tube were projected with such force as to cut the glass face of the mask I had on.' Faraday was learning what to do and what not to do in the laboratory. He was also attending lectures and assisting Davy and the other lecturers in their preparation, sending sharply observed letters to his friends comparing the virtues and vices of the different performers.

A few months after Faraday had started his new employment at the Royal Institution, Humphry Davy made him an offer that would change his life. Davy, as one of Europe's most eminent men of science, had been given a special passport by Napoleon to visit the Continent for a Grand Tour, despite the fact that England was at war with France. He invited Faraday to come with him. Faraday was now faced with an extremely difficult decision. He was, after all, the Royal Institution's employee, not Davy's. If he accompanied Davy on his Tour he would be leaving the post he had worked so hard to secure. On the other hand, if Davy went without him, that would be the end of his philosophical apprenticeship. By following Davy to France and beyond, Faraday would have the opportunity to continue assisting Davy with his experiments as well as meeting a host of European savants. There was also, of course, the fear of the unknown to contend with. Faraday had never travelled more than twelve miles from his home before. The journey to the coast and the sea voyage across the Channel seemed like very daunting prospects to the young Londoner. Nevertheless, despite some last minute misgivings, Faraday decided to go. The party, consisting of Davy, Lady Jane his wife, Faraday and Lady Jane's maidservant, embarked from Plymouth for Morlaix on 17 October 1813.

Faraday was almost immediately sorry that he had come. Problems began when Davy's valet decided at the last minute not to accompany his master on his travels. Rather than wait to find a suitable replacement, Davy asked Faraday to perform some of the necessary services and Faraday, with misgivings, agreed. His misgivings were soon justified, as his performance of a servant's duties led to ambiguities within the travelling party as to just what his social status was. His main troubles were with Davy's wife, Lady Jane. 'She is haughty & proud to an excessive degree and delights in making her inferiors feel her power', he complained in letters home. 'When I first left England, unused as I was to high life & to politeness, unversed as I was in the art of expressing sentiments I did not feel, I was little suited to come within the observation and under the power to some degree of one whose whole life consists of forms, etiquette & manners. I believed at that time that she hated me and her evil disposition made her endeavour to thwart me in all my views & to debase me in my occupation.' Lady Jane was giving Faraday a crash course in the niceties of the English class system. Davy, regarding Faraday as an apprentice, treated him as an apprentice should be treated by his master. Lady Jane, on the other hand, unfamiliar with the codes of conduct that regulated such work-place relationships, saw Faraday carrying out a servant's duties and treated him accordingly. This was the first of many lessons Faraday was to receive during his early career concerning the importance of proper behaviour and the consequences of breaching unspoken codes of conduct. It was an issue on which he remained sensitive throughout his life.

In the meantime, however, Faraday had the French to deal with. His first view of France was 'not at all calculated to impress a stranger with a high opinion of the country', though he did record in his journal his admiration for the country's pigs. He was not much more impressed on arriving in Paris: 'The streets of Paris are paved with equality – that is to say, no difference is made in them between men and beasts, and no part of the street is appropriated to either.' Soon, however, he and Davy were in the swing of scientific affairs as the Parisian savants lined up to pay homage to the English chemist. One such visit provided Faraday with an unrivalled opportunity to see the process of experimental discovery in action. A group of French natural philosophers (Ampère, Clément and Desormes) showed Davy a peculiar new substance that had recently been produced by a French saltpetre manufacturer. The substance had the unusual property of turning into a deep violet vapour when heated. Davy spent the rest of the day – and indeed much of the rest of their stay in Paris – experimenting on the new substance (iodine) in an effort to discover what chemical elements it was made of. After long controversy – during which Davy took every advantage of his mastery of electricity – it emerged that iodine was itself a chemical element, closely related to chlorine – another element that Davy had been responsible for isolating.

Leaving Paris, Davy's party travelled through the south of France, stopping at Montpellier where they were entertained by local philosophers, and passing on into the Italian states. Davy continued his work on iodine as they travelled. By 21 February 1814 they were in Florence, where Davy took Faraday to visit the famous Accademia del Cimento and where he saw Galileo's first telescope. At Florence they carried out a number of experiments on the combustion of diamond in oxygen, using the Grand Duke of Tuscany's great burning-glass. After several attempts, 'Sir H. Davy observed the diamond to burn visibly, and when removed from the focus it was found to be in a state of active and rapid combustion. The diamond glowed brilliantly with a scarlet light inclining to purple, and when placed in the dark continued to burn for about four minutes.' It was, as Faraday remarked, a phenomenon that had 'never been observed before'. As with Davy's experiments on iodine, Faraday was getting his opportunity to watch the process of experimental natural philosophy in action. He was learning how to behave in a laboratory – how to observe, how to take notes and how to manipulate the apparatus. He was also learning about the culture of natural philosophy – how its practitioners behaved, how their institutions worked and how they dealt with one another. In a letter to his mother he described the Accademia del Cimento as 'an inexhaustible fund of entertainment and improvement'.

The next stop was Rome, where Faraday arrived on 7 April 1814. While they were travelling, news was reaching them concerning the progress of the war. A week after arriving in Rome they eventually heard that Allied troops had entered Paris on 31 March. Napoleon had fallen. As elsewhere, the party's time in Rome was a mixture of sightseeing, entertaining and being entertained by the local savants, and carrying out experiments. In his journal, Faraday described in detail a series of experiments he and Davy had carried out attempting to magnetise a needle by means of the solar light, as had been successfully carried out by 'Signor Morrichini of Rome'. After about a month in Rome they left for Naples, and a week later were at the foot of Mount Vesuvius. They climbed the mountain on two successive days, and on the second ascent celebrated with a picnic in the crater: 'Cloths were laid on the smoking lava, and bread, chickens, turkey, cheese, wine, water, and eggs roasted on the mountain, brought forth, and a species of dinner taken at this place ... After having eaten and drunk, Old England was toasted, and "God save the King" and "Rule, Britannia" sung.'

The Grand Tour continued through Italy, into the German states and back to Italy throughout the rest of 1814 and early 1815. One highlight occurred when they passed through Milan and met Alessandro Volta – 'an hale elderly man, bearing the red ribbon, and very free in conversation'. Davy wanted to continue the Tour with a visit to the Ottoman Empire. He changed his mind, however, when he realised that he and the entire party would then have to spend several months in quarantine as a result. More seriously, while Davy and his party were back in Rome for a second visit, news reached them that Napoleon had escaped exile on Elba and had returned to France. In the face of the political turmoil that followed his return, Davy decided that discretion was the better part of valour. On 16 April, Faraday wrote to his mother that they were on their way home, and by 23 April they were back in London. For Faraday the trip had been a life-changing experience. He had spent eighteen months being tutored in the art of experiment by one of its foremost exponents. He had come into contact with a host of eminent continental natural philosophers. The European tour had also brought home to Faraday that if he wanted to succeed in the genteel surroundings of the Royal Institution he would have to learn how to carefully define his relationship with others around him. Experimental dexterity in the laboratory would not be enough to make a natural philosopher. Lady Jane had taught him that he needed to learn the right manners as well.

· Chapter 6 ·

Electromagnetic Rotations

Back in London, Faraday was in a difficult position. Having resigned his position at the Royal Institution to follow his master through Europe, he was now out of work. Bruised by some of his experiences during his travels, he was not entirely sure either that he wanted to continue his career in natural philosophy. He considered returning to his former trade as a bookbinder, where at least he would occupy an unambiguous social niche. As a neophyte in science, Faraday had thought the pursuit of knowledge would make its practitioners 'amiable and liberal' and was full of admiration for 'the superior moral feelings of philosophic men'. Now he was not so sure. Nevertheless, a month after their return, when Davy persuaded the Royal Institution's managers to offer him his old job back with a five shilling wage increase, Faraday accepted. Stephen Slatter, a former assistant porter at the Institution who had been promoted as Faraday's replacement, was demoted to his previous post to make room for the philosopher's apprentice's return. Blackett Thomas Wallis, who in turn had replaced Slatter as assistant porter, lost his job. Davy, in the meantime, had joined the Royal Institution's Board of Managers and had also become one of its Vice Presidents.

Davy now also seemed to think that his chemical assistant was ready for a little autonomy. Not long after his reinstatement at the Royal Institution, Faraday published his first paper in the *Quarterly Journal of Science,* edited by the Institution's new Professor of Chemistry, William Brande. Other papers followed as the chemical assistant gained confidence and Davy gave him more opportunities to work independently. He was still clearly working under Davy's tutelage and direction, however. His publications invariably emphasised that he had carried out this or that piece of chemical analysis at Davy's request. Davy himself also now acknowledged his assistant's contribution to his own experimental researches. Publishing his researches on phosphorus in the Royal Society's *Philosophical Transactions* for 1818, for example, he emphasised how 'in these experiments ... I received much useful assistance from Mr. Faraday, of the Royal Institution; and much of their value, if they shall be found to possess any, will be owing to his accuracy and steadiness of manipulation.' This was praise indeed. Faraday was also preparing himself for a lecturing career. He even took lessons in elocution from Benjamin Smart, one of London's most prominent teachers. This was not a minor undertaking. At half a guinea a lesson, this was a significant investment for Faraday. He was well aware, however, that acquiring the kinds of skills at self-presentation that Smart could teach him would be vital to his success when he first appeared before the Royal Institution's stylish audience. Public speaking was an art as difficult to master as the art of successfully manipulating the apparatus in the Royal Institution's laboratory.

In 1821, Richard Phillips, one of Faraday's old friends from the City Philosophical Society and the editor of the *Annals of Philosophy*, gave him a commission that would put Faraday on the road to scientific greatness, though it would also cause him a great deal of temporary misery. A year earlier, the Danish natural philosopher Hans Christian Oersted had stunned the philosophical world with his announcement of the discovery of the long-sought-for link between electricity and magnetism. Many early 19th-century natural philosophers, particularly those from the German lands, were strong exponents of the fundamental unity of all natural forces. Faraday's master, Davy, himself had sympathies in that direction. Natural philosophers like these were convinced that there must therefore be a link between electricity and magnetism – and Oersted had found it. He had discovered that a magnetised needle, held near a wire carrying an electric current, was deflected. Phillips asked Faraday to prepare a 'Historical Sketch of Electromagnetism' for the *Annals of Philosophy*, outlining the rapid developments made in electromagnetic experimentation since Oersted's stunning discovery. Diligent as ever, Faraday set out to reproduce every experiment published on the matter as part of his research for the historical sketch. With this completed, he carried on experimenting and in September 1821 succeeded in making a current-carrying wire rotate about a magnet. Flushed with success, he quickly published his results in the *Quarterly Journal of Science* and promptly found himself in deep water.

Within days of his paper's publication, Faraday discovered that rumours were circulating questioning his rights of priority to the new experiment. Some months before Faraday had started his own experiments in electromagnetism, William Hyde Wollaston, a friend of Davy's and an eminent natural philosopher in his own right, had carried out some experiments at the Royal Institution to try to make a current-carrying wire rotate around its own axis. He had also made public his views concerning the possibility of electromagnetic rotations in general. It was now suggested that Faraday had stolen Wollaston's work. Faraday was dismayed. This was the kind of accusation that could destroy his career in natural philosophy before it had even started. 'I hear every day more and more of those sounds which though only whispers to me are I suspect spoken aloud amongst scientific men and which as they in part affect my honour and honesty I am anxious to do away with or at least to prove erroneous in those parts which are dishonourable to me', he complained bitterly. As Faraday himself summarised the rumours: 'I am charged 1 with not acknowledging the information I received in assisting Sir H. Davy in his experiments on this subject 2 with concealing the theory and views of Dr. Wollaston 3 with taking the subject whilst Dr. Wollaston was at work on it and 4 with dishonourably taking Dr. Wollastons thoughts and pursuing them without acknowledgement to the results I have brought out.'

Faraday wrote frantically to Wollaston, begging for an interview in which he could defend himself against the accusations. He was well aware that of the charges laid against him, the last – that of acting dishonourably – was by far the most serious. In early 19th-century London, ungentlemanly behaviour was unphilosophical behaviour too. If he were found to have behaved in a manner inappropriate for a gentleman, he would have shown himself unsuitable to be a natural philosopher as well. Wollaston did little to put the distraught chemical assistant out of his misery. 'As to the opinions which others may have of your conduct', he wrote

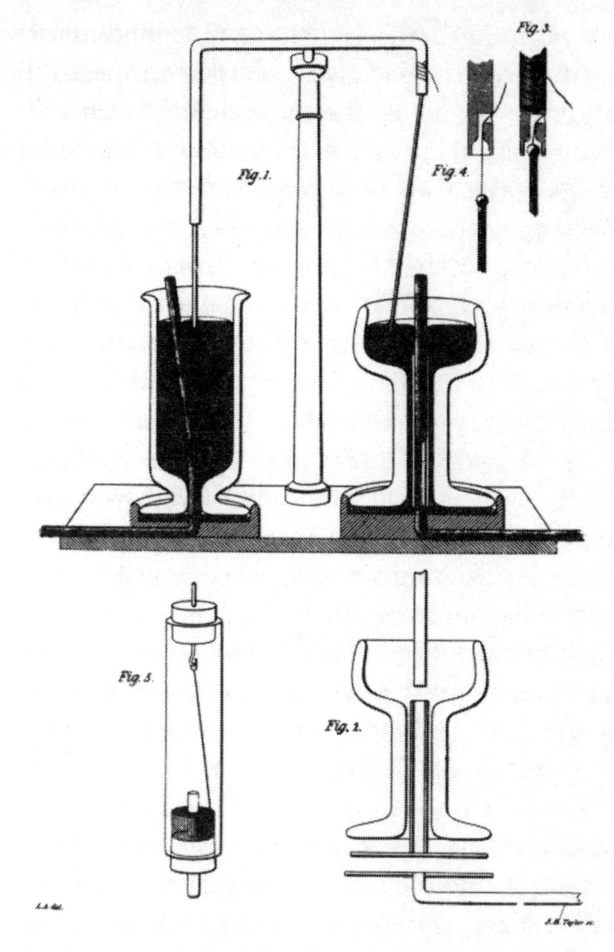

Illustration 4: Michael Faraday's 1820 experiment showing that a wire carrying an electrical current could be made to rotate around a magnet and vice versa. This is the experiment that almost ended Faraday's philosophical career when he was accused of stealing William Hyde Wollaston's scientific ideas.

in reply to Faraday's plea, 'that is your concern and not mine; and if you fully acquit yourself of making any incorrect use of the suggestions of others, it seems to me that you have no occasion to concern yourself much about the matter'. The patrician Wollaston, who had been Joseph Banks's successor as President of the Royal Society, had no intention of letting Faraday off the hook. Part of the problem was whether or not Faraday was the kind of person who could legitimately expect to be counted as a discoverer of anything at all. How appropriate was it for a laboratory assistant to presume to carry out experiments in a field already owned by other, more exalted figures? And could the fruits of a laboratory assistant's labour even be considered to belong to him rather than to his master? It took time and careful diplomacy on Faraday's part before the discovery was generally acknowledged as his own and, as we shall see, the controversy returned to haunt him only a few years later.

Faraday found himself in similar hot water a few years later over his experimental work on the liquefaction of chlorine (an element that Davy himself had been responsible for isolating, and over which he maintained a strong proprietorial interest). Davy had instructed Faraday to carry out a series of experiments including one in which chlorine was heated in a sealed tube, with the result that the chlorine gas became liquid under the pressure.

Before the results could be published in the *Philosophical Transactions*, Davy, by now himself the President of the Royal Society, insisted that Faraday added a note emphasising that the work had been carried out under his instructions and that the result was what he had anticipated. Again, it was a question of property and propriety. Faraday, not unreasonably, clearly felt that his philosophical apprenticeship was by now over and that the fruits of his labour were his own to do with as he pleased. Davy just as clearly did not, and just as a master craftsman considered an apprentice's work to belong to him, so Davy felt that Faraday's experiments were Davy's own intellectual property. Fundamentally, Faraday was finding it very hard to find a niche for himself in the scientific world. He did not yet know the rules of the game or how to behave properly.

· CHAPTER 7 ·

F.R.S.

In 1823, just ten years after Faraday had embarked on his philosophical apprenticeship, he was nominated a Fellow of the Royal Society. He had come a long way from being a humble journeyman bookbinder. Whatever Davy felt on the matter, Faraday was by now widely recognised as a distinguished natural philosopher in his own right – particularly as a result of his contested discovery of electromagnetic rotations. His continental travels with Davy had stood him in good stead and he was in regular correspondence with some of the great philosophical names – like André-Marie Ampère and Charles-Gaspard de la Rive – whom he had met during the Grand Tour. He had been elected to membership of august scientific organisations such as the Cambridge Philosophical Society. When the notice of his nomination to Fellowship of the Royal Society was posted, however, it drew an angry response from some quarters. A coterie of Fellows, friends of the former President, William Hyde Wollaston, threatened to blackball Faraday. They still felt that Faraday's behaviour over the discovery of electromagnetic rotations had shown him to be a blackguard and a cad. They certainly felt that he had shown himself unfit to be permitted to join a society of gentleman philosophers. Faraday had to work hard to mollify Henry Warburton, the leader of the pro-Wollaston clique, and convince him of his honourable intentions.

Far more devastating for Faraday, however, was Davy's opposition. Davy was furious when he found out that Faraday had allowed himself to be nominated for Fellowship. He promptly insisted that his chemical assistant withdraw his name from consideration. Faraday's response was that he could not do so without offending the long list of Fellows who had signed the nomination. The reasons for Davy's vehement opposition to Faraday's Fellowship have been a focus for much historical speculation. The most commonly cited reason is sheer envy. According to this view, Davy was simply jealous of his erstwhile protégé's success and wanted to stand in his way. This is extremely unlikely (though it may well have been how Faraday himself regarded the matter). Davy at this time was himself President of the Royal Society, after all. He was widely celebrated as England's greatest chemist and one of the greatest in Europe. Faraday, on the other hand, had a handful of scientific publications to his name – some of them, as we have seen, distinctly controversial. Faraday's star was certainly on the rise, but it was very far indeed from eclipsing Davy. Other historians have cast the relationship between the two as a classic father–son conflict, with Davy as the father fearful of being upstaged by Faraday the rebellious son.

In fact, the reason for Davy's opposition to Faraday's nomination to a Fellowship can be found in the circumstances surrounding Davy's own embattled Presidency of the Royal Society. The Royal Society throughout the 1820s was riven by conflict between opposing camps. On one side were followers of the tyrannical former President, Sir Joseph Banks, who had remained President of the Society until his death in 1820. On the other were a group of reformers, led by the mathematician Charles Babbage and the astronomer John Herschel, who wanted entry into the Royal Society to be based on merit rather than nepotism. Davy's selection as President had been very much a compromise between these two camps. His proposers hoped that he would be patrician enough to please the Old Guard, and radical and scientifically eminent enough to mollify the young Turks. In reality, of course, Davy as President failed to satisfy anybody – a fact of which he was all too painfully aware. In these circumstances, Davy regarded Faraday's nomination as a calculated insult. It was an effort by his enemies to undermine his Presidency by making him appear foolish and guilty of nepotism himself. In the event, despite his opposition, Davy did nothing further to stand in the way of Faraday's election – which as President he could certainly have done had he wanted. Faraday was duly elected a Fellow of the Royal Society on 8 January 1824.

Despite his anger concerning Faraday's disloyal (as he saw it) behaviour over the Fellowship election, Davy continued to support his protégé. A year later, he proposed to the Royal Institution's Board of Managers that Faraday be promoted to a new position more suitable to his rank as a Fellow of the Royal Society and an eminent natural philosopher in his own right. The Managers duly resolved that 'Mr. Faraday be appointed Director of the Laboratory, under the superintendance of the Professor of Chemistry'. As Director of the Laboratory, Faraday now had considerable freedom to pursue his own experimental inquiries as he wished. He was no longer at Davy's or his successor William Brande's constant beck and call. He was his own man at last. Faraday did, however, find himself increasingly involved in the day-to-day running of the Royal Institution, particularly as it found itself in increasing financial difficulties during the second half of the 1820s. The halcyon days of the Royal Institution's early years were by now well and truly over. Lectures were no longer drawing in the fashionable crowds in the same way, particularly now that the dashing and handsome Sir Humphry was no longer a regular feature on the billing. Something needed to be done to restore the Royal Institution's fortunes, and Faraday was the man charged with the task of doing it.

During the course of 1825, Faraday gradually introduced a new feature into the Royal Institution's range of provisions for its members and subscribers. On Friday evenings, members and their guests were invited to come to the Institution to view exhibits laid out on display in the Library and to hear Faraday or an invited speaker give an account of the latest scientific discovery or curiosity. These Friday Evening Discourses, as they came to be called, were to become crucial features of the Institution's offerings. They also came to play a crucial role in the making of Michael Faraday as the metropolis's premier interpreter of nature for the leisured classes. They were certainly an almost immediate hit with the Royal Institution's genteel clientele. Faraday also reorganised the Royal Institution's regular lecture courses. As he explained to the Board of Managers, 'According to the present arrangements of the Institution, the lecture Season usually commences in the beginning of February and continues till the month of June including a period of about four months. Of course all who subscribe to the lectures generally have the opportunity of attending the whole of the lectures during the Season, but the fashionable Season in London seldom commences until after Easter, so that the persons who come to town at that time although they pay the same subscription have only half the advantage of them who are in London for the whole period.' Faraday's proposal to adjust the lecturing schedule to conform to the rhythms of fashionable metropolitan society had a major impact on the Royal Institution's financial fortunes.

By the end of the 1820s, Faraday's philosophical apprenticeship was well and truly over. For one thing, his old master Sir Humphry Davy was dead. Davy died alone in Geneva in 1829. Describing his relationship with Davy after his death, Faraday remarked that 'I was by no means in the same relation as to scientific communication with Sir Humphry Davy after I became a Fellow of the Royal Society as before that period'. That Fellowship election can be taken as the point when Faraday cut himself loose from the bonds of patronage that linked him to his master, and became his own man. Whether he liked it or not (and there is a fair amount of evidence that he didn't), Faraday had learned and gained a great deal from his master. He had learned how to become a superlative experimenter. He had learned the importance of public performance. He had been given the opportunity to rub shoulders with some of the greatest European natural philosophers. He had been provided with a secure institutional niche with unparalleled resources. But Faraday was also adept at self-fashioning. The trials and tribulations surrounding his relationship with Lady Jane, the electro-magnetic rotation fiasco, and his troubled nomination to Fellowship of the Royal Society had made him highly sensitive to his social position. He knew that he had to step very carefully indeed and present a very carefully contrived face to the world if a working-class lad like him was to make a name for himself as a man of science in a society of gentlemen.

Part III: Radical Electricity

· Chapter 8 ·

A Dangerous Science

Two of Faraday's 19th-century biographers – Henry Bence Jones and J.H. Gladstone – mention how, at the peak of his career, he found himself outed as an apparent supporter of dangerously radical and heterodox theories about the relationship between electricity and life. Gladstone revealed that he himself had heard the slander propagated by 'an infidel lecturer on Paddington Green'. Henry Bence Jones, the author of Faraday's *Life and Letters*, made a similar reference, also to a lecturer (quite possibly the same one) on Paddington Green. The lecturer, a Mr Wild, had been disputing the Old Testament account of creation and had cited experiments by Faraday as evidence for his claims. The experiments, carried out before audiences at Oxford, Cambridge and London, had demonstrated the electrical nature of life by producing animalcules and maggots by electrical agency. Faraday had underlined his experiments with the remark to his audience that: 'Gentlemen, there is life, and, for aught I can tell, man was so created.' He inferred from his experiments that man could be created, and in all probability was created, in the same way as by his experiments. According to the 'infidel lecturer' Mr Wild, 'so unpalatable were [Faraday's] views, and contrary to what was received as orthodox, that the authorities under whose auspices the lectures were given ... had them discontinued'.

Both biographers mentioned this scurrilous rumour only to dismiss it. The man who was by then widely recognised as the doyen of British, if not European, electricians would never have made such an outrageous suggestion. The episode is very revealing, nevertheless, of some of the sensitivities surrounding the science in which Faraday was to make his reputation. From the late 18th century onwards, electricity was, in many ways, a politically suspect science. It was the science of atheists, materialists, political radicals and revolutionaries. According to some of its critics, electricity was even to blame for the French Revolution. It was relatively common in 18th-century electrical circles to argue for some connection between electricity and the *vis nervosa* – the stuff of life. In some posh *salons*, a good dose of electricity was widely recognised as the perfect antidote to infertility and impotence. Taking this argument to its extreme, materialists argued that the connection between electricity and the *vis nervosa* showed that life was just matter in motion after all – and human beings just machines. This was a dangerous argument. It meant no God, no Church, no divinely ordained social hierarchy. In short, it meant the Rights of Man. When the English radical Joseph Priestley trumpeted that 'the English hierarchy, if there be anything unsound in its constitution, has reason to tremble even at an air-pump or an electrical machine', this is the sort of thing he had in mind.

By the beginning of the 19th century there certainly seemed to be plenty of evidence to support this view. The galvanic, or voltaic, battery – one of the period's experimenters' main tools of philosophical investigation – was the product of the dispute between Luigi Galvani and Alessandro Volta over the existence of animal electricity. Volta – the battery's inventor – was convinced his instrument showed that animal electricity was nonsense. Not everyone agreed. Giovanni Aldini, Galvani's nephew, visited London in 1803 to defend his uncle's reputation and the doctrine of animal electricity. On 17 January he was given the opportunity of experimenting on a human subject – Foster – a murderer executed at Newgate. Aldini connected his subject to the poles of a large galvanic battery. Electricity was passed between the dead man's ears, between his mouth and his ears and between his anus and his ears. The result was a startling exhibition of contractions and convulsions. 'On the first application of the process to the face, the jaw of the deceased criminal began to quiver, the adjoining muscles were horribly contorted, and one eye was actually opened. In the subsequent part of the process the right hand was raised and clenched, and the legs and thighs were set in motion. It appeared to the uninformed part of the by-standers as if the wretched man were on the eve of being restored to life.' Aldini's experiments were at the time a minor sensation. They were widely reported in the press and patronised by the fashionable, even including the Prince of Wales.

Andrew Ure's experiments, fifteen years or so later, fitted into the same mould. He experimented in 1818 on the body of Clydesdale, executed for murder in Glasgow. Like Aldini, he hooked the dead man up to a galvanic battery and passed a current through different parts of his body. Extraordinary effects were produced when the corpse was electrified: 'every muscle in his countenance was simultaneously thrown into fearful action; rage, horror, despair, anguish, and ghastly smiles, united their hideous expression in the murderer's face, surpassing by far the wildest representations of a Fuseli or a Kean. At this time several of the spectators were forced to leave the apartment from terror or sickness, and one gentleman fainted.' In these experiments, the criminal's body was transformed by electricity into something like an automaton, reproducing at the experimenter's will the stylised gestures and grimaces of Regency art and theatre. In the absence of the mind's power controlling the body's movements, electricity could be used instead to simulate animation. Ure's experiments were even published in the *Quarterly Journal of Science*, in-house journal of the Royal Institution. Ure was to gain notoriety in the 1830s as the author of the *Philosophy of Manufactures*, a paean to the factory system which celebrated the way that the introduction of automatic machinery meant that workers could be treated like automata too.

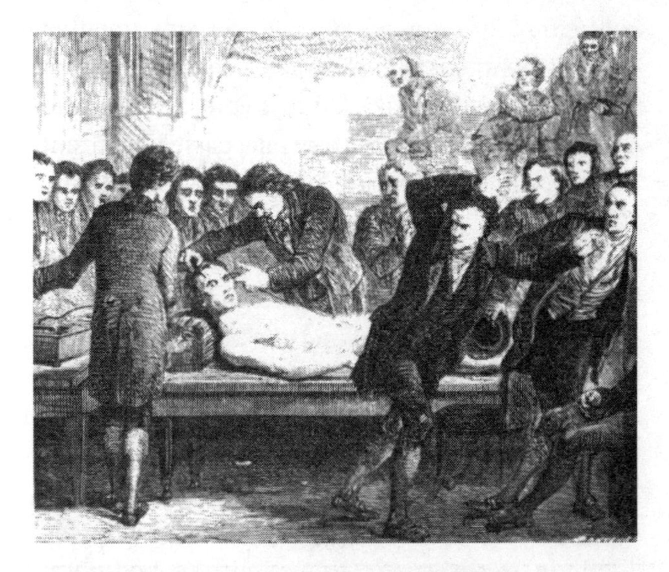

Illustration 5: Andrew Ure performing galvanic experiments on the corpse of the executed murderer, Clydesdale. The experience was clearly too much for some of the witnesses fleeing the scene.

In the same year that Ure published his experiments in the *Quarterly Journal*, 1818, Mary Shelley published *Frankenstein*, placing the problem of artificial life at the novel's heart. Shelley's story of a student discovering the secret of life, stealing body parts to produce his monster, bringing it to life and fleeing in terror from his creation, leaving it to wreak its path of devastation, is deeply ingrained in modern culture. Less clear, maybe, is the extent to which Mary Shelley was drawing on some of the most cutting-edge contemporary discussions in natural philosophy in putting her tale together. Aldini's and Ure's experiments show that there was nothing implausible about the possibility of creating artificial life through electricity, at least in some early 19th-century circles. Mary and Percy Bysshe Shelley were friends of the radical doctor William Lawrence, who was embroiled in bitter controversy with the physician John Abernethy over just this issue – whether life was simply a product of material organisation or whether it was something superadded to matter. Lawrence would end up condemned for blasphemy over his claim that life was simply a matter of organisation. Discussions of the links between electricity and living matter were just as contentious and politically suspect during the early decades of the 19th century as they had been in the closing years of the 18th century. This was the filthy underbelly of electrical science, far removed from Faraday's Royal Institution. It was what made electricity look to some like a risky prospect and some of its practitioners mad, bad and dangerous to know. If Faraday wanted his favourite science to be fully respectable then he had to rid the world of natural philosophy of these galvanic subversives.

· CHAPTER 9 ·

THE WIZARD OF FYNE COURT

In 1836, some experiments carried out by an obscure West Country squire, Andrew Crosse, at his home, Fyne Court near Broomfield in Somerset, seemed to make the creation of life from dead matter through electricity into a reality. Crosse was the son of a family steeped in radical politics, and was proud of the family tradition that his father had been among the first to raise the tricolour over the ruins of the Bastille during the French Revolution. He was also an enthusiast for electrical experiments and had devoted himself to producing them on a massive scale. The trees in the grounds of Fyne Court were festooned with wires which Crosse used to try to collect lightning for his experiments. He was feared locally as the 'thunder and lightning man'. Crosse's main interest was the production of artificial crystals by means of electricity. With this end in mind, he had constructed massive banks of voltaic batteries that passed currents of low intensity through various chemical solutions over a period of months. These experiments had only recently brought Crosse to public attention when he visited the 1836 meeting of the British Association for the Advancement of Science held that year in nearby Bristol. There, he had been hailed as the epitome of the isolated, self-effacing natural philosopher, interrogating nature without thought for private gain.

Returning to Fyne Court, Crosse had continued his experiments. In one of his experimental set-ups, a dilute solution of silicate of potash saturated with muriatic acid was slowly dripped over 'a piece of somewhat porous red oxide of iron from Vesuvius'. Two platinum wires from the poles of a voltaic battery were connected to this piece of volcanic rock so that the stone was kept constantly electrified as the fluid was dripped over it. Crosse hoped that this arrangement would produce artificial crystals of silica. The actual results, therefore, were entirely unexpected: 'At the end of fourteen days, two or three very minute specks or nipples were visible on the surface of the stone, between the two wires, by means of a lens. On the eighteenth day these nipples elongated and were covered with fine filaments. On the twenty-second day their size and elongation increased, and on the twenty-sixth day each figure assumed the form of a perfect insect, standing on a few bristles which formed its tail. On the twenty-eighth day these insects moved their legs, and in the course of a few days more, detached themselves from the stone, and moved over its surface at pleasure, although in general they

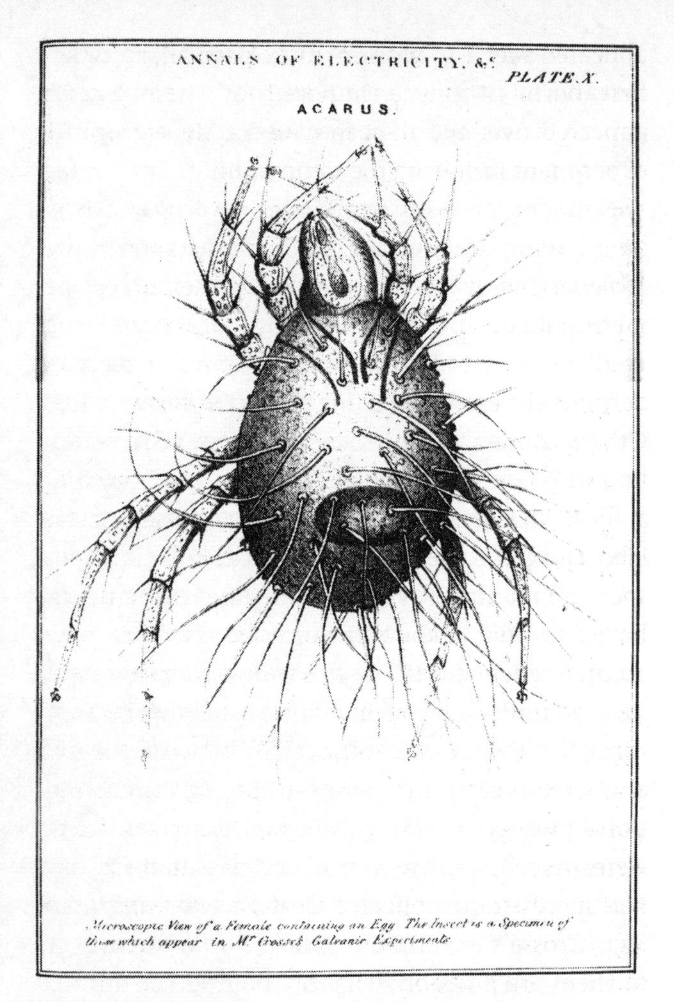

Illustration 6: A drawing of one of Andrew Crosse's notorious electrical insects, the *acarus crossii*. It caused major controversy when a number of these insects crawled out of one of Crosse's electrical experiments in 1836.

appeared averse to motion, more particularly when first born.' About a hundred of these insects appeared over the next few weeks. Repeating the experiment produced the same result.

News of Crosse's experiments spread rapidly after having been leaked to a local newspaper, the *Somerset County Gazette,* and then picked up by the metropolitan and national press. Reactions were predictably mixed. According to some, Crosse had definitively established the material basis of life. Others condemned him for blasphemously meddling with God's laws. There was even a rumour that a local vicar had attempted an exorcism of his laboratory. Crosse himself was carefully circumspect about the way in which the insects might have appeared, and never suggested that they were an original creation. Samples were sent to eminent natural philosophers, including Michael Faraday himself at the Royal Institution, for their inspection and were even put on show at one of the Institution's prestigious Friday Evening Discourses. There were, indeed, some reports in the press that Faraday had successfully replicated Crosse's experiment. In fact, Crosse's experiments and Faraday's response to them are probably what lay behind the 'infidel lecturer' and his claim that Faraday had produced life from electricity. Faraday, of course, had done no such thing. Others, however, did take Crosse's work seriously and attempted to repeat it. The electrician

W.H. Weekes reported to the London Electrical Society (of which Crosse was also a member) that he had successfully repeated Crosse's experiments, producing insects from electricity.

A short story – 'The New Frankenstein' – in the scurrilous *Fraser's Magazine* for 1838 offers a good example of the way Crosse's experiment was used to highlight the dangers of materialism, as well as to poke fun at the pretensions of electrical experimenters. The anonymous tale, written in the style of E.T.A. Hoffmann's gothic fantasies, featured a German student as hero. Like the original Frankenstein, he was obsessed by natural philosophy and the prospect of creating life by artificial means. Studying in Paris, he carried out experiments that were a clear parody of Andrew Crosse's crystal- and insect-making activities: 'I was the first to discern that chrystals [*sic*] are to be produced by the galvanic battery, and animal life from acids ... I got some volcanic dust from Etna, which I impasted with muriatic acid, and after a while distinguished, though inaudible save with an ear trumpet – or thought I could distinguish – a *hum*, like that of fermentation. What was my delight to find that there was vitality in the mass – that these atoms daily grew in size! They were of the *bug* species; not unresembling what the French call a *panaise*.' His researches eventually led him to the rediscovery of Frankenstein's monster.

On examining the monster the intrepid student found that, while he was physically sound, he lacked animation. The monster 'had only a talismanic existence – was a mere automaton – a machine – a plant without the faculty of motion'. He lacked the distinctive vitality of the human body: 'there was a mechanical trepidation of all his fibres; his nerves quivered, but not with sensibility, and his whole frame had a convulsive motion; whilst his head moved from left to right and right to left, like that of a Chinese mandarin.' Electricity, with which the monster was sufficiently charged to have struck the student to the ground as if from a bolt of lightning when he first approached him, could provide the body with only the semblance of life. Resolving to remedy this defect, the student travelled Europe, stealing the minds of a roll-call of the Continent's greatest thinkers – Goethe, Schelling, Shelley, Coleridge – and distilling them into the monster's brains with a pair of Perkins's Tractors – a well-known item of contemporary electrical quackery. The result, unsurprisingly, was that the monster went insane. The story concluded as the student hero was swallowed up into hell while trying to provide his creation with a soul.

The story certainly took the mickey out of radical electricity. Its identification of the hero as a German student would have reminded its readers of the excesses of early 19th-century German Romantic

Naturphilosophie, with its wild talk of the world soul and the cosmic animal. The tongue-in-cheek list of great European minds raided to provide the monster with intellect would have done the same thing. Paris, where the student studied, was, of course, well known to *Foster's Magazine*'s readers as the fount of all atheistical and materialist gibberish. The comparison of the monster to a Chinese mandarin pandered to English prejudices, which viewed foreigners in general – and Orientals in particular – as machine-like in their subservience to despotic government. This comparison of the monster to a machine, and the inability of electricity to provide it with an authentic life of its own, is the key to the parable's serious message. The tale's moral was unambiguous. Producing life by electricity was nothing better than a parody of nature. Even if the galvanic power was sufficient to simulate the machine-like physical properties of life, it could not hope to reproduce the moral and the spiritual. Any attempt to do so would, in any case, be little short of blasphemy. Machinery's animation, the story implied, was a fraud. Frankenstein's monster could indeed be restored to life by galvanic agency. It could even be provided with a rational mind. The galvanic life, however, was only an illusion, a simulacrum of reality. Frankenstein's monster had no soul and could not therefore be truly living.

Andrew Crosse's production of his living *acarus crossii* (as they were soon baptised) from inanimate matter through electricity left orthodox gentlemen of science in something of a quandary. It was difficult to dismiss Crosse as a charlatan. After all, only a few months before the *acarii* started crawling out of his electrical apparatus, he had been lionised on stage at the Bristol meeting of the BAAS as the ideal gentleman natural philosopher. He could not now very easily be denounced as a fraud by the very men who had been responsible for setting him on a pedestal in the first place. He was also, in any case, a gentleman. Instead, by and large, they tried to find ways of sanitising his discovery without dismissing it outright. The geologist and Anglican divine, William Buckland, for example, suggested that rather than having created life from dead matter, the electricity had revivified fossilised insect eggs embedded in the volcanic rock that Crosse had used for his experiments. Such an explanation had the virtue of saving Crosse's reputation while simultaneously rejecting the materialist gloss that otherwise seemed inevitable. We will discover Faraday's own response in the next chapter. Materialists and radicals, on the other hand, seized on Crosse's experiments with alacrity. They were proof positive of the identity of electricity and life – and therefore of the need for urgent social revolution.

· Chapter 10 ·

The Electrical Universe

One political radical during the 1830s who made full use of Andrew Crosse and his *acarii* was Thomas Simmons Mackintosh. The former factory manager and ardent follower of the utopian socialist Robert Owen put electricity to work as the foundation of a revisionist grand theory of the natural order. His 'electrical theory of the universe' was designed as down-to-earth science for the common man that would remove 'mystery' from the study and understanding of the natural world. In articles appearing in both the *Mechanic's Magazine* and the Owenite *New Moral World*, Mackintosh gave detailed accounts of his theory, showing how electricity was the power through which the universe maintained itself in motion. This was a political project. Doing away with explanations of the natural order based on 'mysterious' agencies was central to establishing an enlightened social order: 'We do not hope to convince those who are stubbornly bent upon explaining the mysteries of nature, by introducing other mysteries still more incomprehensible than those which they profess to explain. We should not have entered upon the subject were it not that we are firmly persuaded that before we can hope to elevate the physical and moral condition of man, we must carefully and minutely investigate the laws by which his physical nature is governed, and thereby we shall obtain a sound basis for our moral structure.' Electricity was to be a means of dispelling hidden and occult causes and of placing the happiness of mankind, which was the proper aim of all physical science, on a proper footing.

Mackintosh's theory was part of the repertoire of lectures he delivered at Halls of Science in towns and cities throughout the Midlands and the north of England. It played its role in Owenite campaigns to re-found science on rational, socialist principles. By all accounts, his lectures were extremely popular. A correspondent in Liverpool enthused that 'the attendance at the institution here is on the increase, in consequence of Mr Simmons Mackintosh, of London, having commenced a course of six lectures on the "Electrical Theory of the Universe"'. At Ashton a month or so later, it was reported that one of Mackintosh's lectures 'was better attended than any previous lecture in this place'. At Salford, his lecture 'on the attractive and repulsive moral forces ... illustrated by several very humorous drawings' was said to have conveyed 'much instruction and gratification to the audience'. He explained in his lectures how all motion throughout the solar system was the result of electricity. This applied not only to the movements of planetary bodies but to the 'minuter processes of vegetation, oxidation, and vitrification' as well. He drew on a whole range of sources to establish his claims. The Quaker electrician Robert Were Fox's experiments on the electricity of geological structures were invoked to establish the galvanic fluid's role in the Earth's formation. Thomas Pine's work on electro-vegetation established electricity's role in the growth of plants. Andrew Crosse's electrical production of insects played its role in demonstrating the electrical basis of animal vitality. Mackintosh drew heavily on Aldini's and Ure's experiments on human bodies as well, to establish the central role electricity played in maintaining human life. Their work showed how 'the animal system is a bundle of circles, each connected with the others, like the wheels of a watch, or like the different parts of a steam engine, and that the primary circle, the main spring, which may be said to originate the animal functions, is the nervous, and that the nervous circle is actuated by electrical agency'.

One crucial consequence of Mackintosh's theory was that it posited a definite and limited time-span for the solar system. The natural electricity of the Earth and other planets was gradually being dissipated. The repulsive power between the Sun and the planets was gradually being weakened, so that eventually they would all fall into the Sun. The planets were in effect 'propelled down an inclined plane by the power of electricity'. In this view, along with the view of man as 'an organised machine', the possibilities of utopia were limited to the terrestrial sphere. Utopia was possible, since, if 'the roots of moral action be in physical organization, in proportion to our knowledge of that organization we can stimulate or retard that action, and turn it into courses which would lead to a larger amount of happiness both to the individual and to the society of which he forms a member'. No organised machine could last for ever, though. As Mackintosh put it: 'the river flows because it is running down; the clock moves because it is running down; the planetary system moves because it is running down; every system, every motion, every process, is progressing towards a point in which it will terminate; and life is a process which only exists by a continual approach towards death. Eternal life and perpetual motion are almost, or altogether, synonymous.' Replacing the Newtonian philosophy with the electrical theory of the universe meant replacing the whole social, political and religious order which underpinned early 19th-century life. If man and the universe were electrical machines, then the Kingdom of Heaven could be founded only on Earth.

A very similar view of the consequences of recognising the human body as an electrical machine came from another radical source. Eliza Sharples, the political radical and mistress of publisher Richard Carlile, made extensive use of electricity in her Isis lectures at the Rotunda on Blackfriars Road during the early 1830s, deploying an electrical model of human physical organisation to underline her anti-clerical message. Sharples had established herself as a lecturer at the Rotunda while Carlile was once more in prison. Her lectures there were widely castigated for their infidel subversive character. The prospect of a woman publicly expounding and popularising such dangerous and heterodox opinions drew prurient comments from the editor of *The Times*, which Sharples herself reprinted in her own periodical *The Isis*, sharply remarking that comments on dress and physical appearance were unlikely to have been deemed relevant had the lecturer in question been male. Her position was rendered even more difficult as it became known that she was carrying Carlile's child and described their relationship as a 'moral marriage'. This blatant transgression of social propriety was too much even for most of the radical press to swallow.

Sharples turned to electricity in her lectures to 'account physically for the moral usefulness of proper education, and show how, why, and to what end, the body may be impressed by an action upon the brain morally'. Her argument rested on the assumption that the human body was a 'self-acting electrical machine, sustained by currents of atmospheric air and liquids'. The brain had the capacity to 'concentrate and direct its electrical power' to any part of the body by the exercise of will, acting just as would a galvanic battery applied to a corpse to produce movements of various kinds. The brain was the fount of all bodily functions, and action on the brain had direct consequences for the rest of the body. The human race could therefore be morally and physically improved by proper education and environment.

Electricity played its role in confounding the established Church as well. A number of Sharples's lectures at the Rotunda were framed as 'Discourses on the Bible', and aimed at providing a materialist exegesis of scriptural texts. Resurrection was reinterpreted as intellectual enlightenment and explicated in terms of electrical activity in the brain: 'It is understood, by physiologists, that the cerebral action which produces mind or knowledge, is literally an action of spirit upon water, or indeed a formation of water on the brain, in the small healthy quantity, by the electrical excitement which is produced in the effort to think and comprehend ... The brain is a self-organizing galvanic pile. Water and spirit, or electricity, we know, are the essential principles of galvanic action. So that it is a literal truth that the mind, the second birth of man, is born of the water and the spirit.' Electricity provided a means of restoring literal meaning to scripture, re-casting superstition as natural philosophy.

For Sharples, the notion of a non-material afterlife was anathema: 'The conceit that spirit can retain an identity without the aid of the body, is that of superstition and madness.' It was crucial to her and Carlile's project that any such misguided notion be dispelled: 'It is important that man should be divested of his conceit of eternal life; for it sets at naught the value of all calculations for the welfare of that which is present.' The rationale was again electrical: 'All electricity depends upon certain arrangements of materials, without which it cannot exist; so that the imagination of life without body, is like the creation of all things out of nothing.' Life and spirit were straightforwardly identified with the electric fluid. Since that fluid could not exist without a material basis, nor could spirit. Recognising electricity as the basis of all life allowed for a reinterpretation of the Gospels as an exhortation to work towards the transformation of moral action during life, rather than to wait for salvation in a hypothetical and physically impossible resurrection.

Views like these were common currency in

materialist and radical circles. They provided an important alternative cosmology that helped validate egalitarian and progressive politics. The flamboyant Francis Maceroni, formerly aide-de-camp to Joachim Murat – dashing cavalry general, King of Naples and Napoleon's brother-in-law – and author of a notorious physical force pamphlet during the Reform debates of the early 1830s, held that all earthly phenomena, from the eruptions of volcanoes to the actions of 'brains, nervous ganglions, and nerves', were the result of the workings of 'real electrical machines'. Electricity played its role in holding the Earth together as 'the great PAN, or whole'. These views carried weight outside radical networks as well. The anonymous and best-selling *Vestiges of the Natural History of Creation*, published in 1844 (and actually written by the respectable Edinburgh publisher Robert Chambers), made use of a version of Mackintosh's electrical theory to explain the progressive unfolding of natural law. Even the highly orthodox Tory Alfred Smee turned to an electrical account of the workings of the human body to confirm Anglican doctrine. According to Smee, the body was a collection of electrical instruments and the brain a battery. Unravelling the brain's complex electrical circuitry showed that proper religious orthodoxy was hard-wired into the brain and that misfits like atheists and Catholics suffered from faulty wiring.

Electricity was a very malleable and versatile ideological resource in early 19th-century Britain – and men like Faraday had to work hard to keep it respectable.

· CHAPTER 11 ·

GALVANIC MEDICINE

Sharples's and Carlile's involvement with electricity was not limited to using it as a foil in their lectures and pamphlets. Electricity could provide a route to practical, physical betterment as well as to the moral improvement of the human race. During the early 1840s, both were involved with the medical galvanist and mesmerist William Hooper Halse. Carlile became increasingly interested in both galvanic medicine and mesmerism during the 1830s. In 1841 he brought Halse to London to give demonstrations of animal magnetism or mesmerism in Chancery Lane. A little later, when Sharples became ill, she visited Halse in Devon for treatment. Galvanism not only provided a rationale for the body's and for society's actions and illnesses, it provided a therapy for physical as much as moral ailments as well. Galvanic medicine was increasingly fashionable in early Victorian England. Medical electricity had been an occasional feature of the therapeutic repertoire of practitioners since the 18th century. During the 1830s, the proliferation of new galvanic technologies, itself partly driven by the needs of a medical market, led to a more widespread availability of the galvanic fluid. Electricians hawking their latest inventions would often target a medical audience. The instrument-maker Edward Clarke, for example, announcing his controversial Magnetic Electrical machine during the mid-30s, was quite emphatic that medical gentlemen would find his invention of great utility. Instrument-makers' catalogues of the period frequently included electrical apparatus adapted for medical purposes in their range of advertised products.

Application of electricity for therapeutic purposes certainly seems to have been a routine activity on the part of electricians. William Sturgeon mentions that he had been called upon to administer electricity to some young men who had fallen through the ice on the river near Woolwich. His efforts were unsuccessful, due, he recorded, to 'two hours having been occupied, by the usual routine of medical treatment, before the batteries were employed'. Andrew Crosse was also reputed to have offered his services as a medical electrician to his less affluent neighbours. The administration of shocks for the purpose of entertainment was also a routine part of electrical performances at popular venues such as the Adelaide Gallery. A number of professional 'medical galvanists' also offered their services to the public. Halse, who eventually settled in London during the early 1840s, was one of the more successful of these practitioners. He had links with at least some members of London's electrical community, having published on several occasions in William Sturgeon's *Annals of Electricity*. In one contribution to Sturgeon's *Annals*, he detailed a number of experiments in which new-born puppies were immersed in water until drowned before being returned to life by means of electricity. He recommended his experiments to the medical profession as proof positive of the 'astonishing powers of galvanism' in supplying the nervous fluid. An abstract of this paper was published in, among other places, the Owenite *New Moral World*.

Having arrived in the metropolis in 1843, Halse spared no effort in publicising his activities. His pamphlets were packed with grateful testimonials and stories of miraculous cures from his days in Devon. He continued to combine galvanism and mesmerism in his repertoire, describing himself in a mesmeric broadsheet as a 'medical galvanist, proprietor of the galvanic family pill'. Potential patients wishing to be galvanised had the choice of either attending at Halse's residence, at the cost of one guinea a week, for half an hour's daily treatment, or purchasing one of his machines for their own use at the price of ten guineas. This, incidentally, was almost twice the price of an electro-magnetic machine or induction coil purchased from an ordinary instrument-maker. One of the most striking features of Halse's self-promotion was the way in which he described his patients' bodies. Simply speaking, the sick body was out of control. It was subject to convulsions, spasms and twitches in much the same way that the corpses electrified by Aldini and Ure had been. Or rather, in this case, the sick body did to itself what could also be done with electricity, and galvanism was the means of *regaining* control. The sufferer's body had become like an automaton divorced from the mind's domination. Halse specified a wide range of complaints for which galvanism was the cure. He listed 'all kinds of nervous disorders, asthma, rheumatism, sciatica, tic doloureux, paralysis, spinal complaints, long-standing headaches, deficiency of nervous energy, deafness, dulness [*sic*] of sight, liver complaints, general debility, indigestion, stiff joints, epilepsy, and recent cases of consumption'.

It was of particular use in cases of nervous disorder, since in such circumstances his galvanic apparatus was 'actually supplying the nerves with their proper stimulus during the whole of the operation'. In almost all these cases, the patient's body was in some way out of control. Mrs S. of Torquay, for example, consulted Halse over the case of her daughter, 'whose face on one side was dreadfully distorted, the right side of the mouth and right eye being so contracted that they almost touched. She could not close her eye either by day or night, and, as may be supposed, it had a very awful appearance.' Halse galvanised her face every day for a month and her face 'resumed its natural appearance'. Electricity had been a means of regaining control over her body. Another example was an Oxford clergyman suffering from indigestion, the result, Halse suggested, of too much studying. 'He was highly nervous, and totally unfit to enjoy society; his eye was dull in the extreme; he was subject to spasms, costiveness, headache, extreme languor, lowness of spirits, very excitable in sleep, feet and legs like ice, and not the least appetite. He frequently saw the vision of a white cat running about the house, and he declared to me that it appeared just as real to him as if it had actually been a cat.' The clergyman had completely lost the ability to regulate his own body and mind. Again, it took a month of daily galvanism before the symptoms, including the white cat, eventually disappeared.

Halse was explicit in his condemnation of the medical profession for being both mercenary and unscientific. Halse's patients had almost invariably suffered at the hands of unscrupulous physicians before being saved by his galvanic ministrations. They were unscientific because they failed to realise the therapeutic efficacy of galvanism and the scientific doctrines upon which its application was based. They were mercenary because they kept on plying their patients with useless and expensive drugs they knew full well would never work. Even galvanism's disrepute as a therapeutic agent could be laid at the door of some of the profession's cack-handed efforts to use it. A large number of them were no more than 'a set of bunglers who have recourse to this agent, and who know no more how to manage it than a donkey knows how to manage a musket'. The possibilities of galvanism as a scientific cure-all opened the door to a new comprehensive understanding of the body and a means of restoring the physically (and the spiritually) sick to their proper place. Medical galvanism – like radical electrical theories – offered a challenge to orthodox views of what counted as being properly scientific. These were challenges that Faraday devoted much of his career to combating. The brand of electrical science that Faraday would sell to his genteel upper-class audience at the Royal Institution would be carefully shorn of any radical content and tailor-made not to offend their sensibilities.

Part IV:
Royal Institution Science

· Chapter 12 ·

Bringing Down the House

By the end of the 1820s, Faraday was firmly ensconced at the Royal Institution. He was the Director of the Laboratory, with a salary of £100 a year. He was provided with rooms at the Institution, including free candles and coal – a not inconsiderable perk, worth as much as half his salary in kind. In 1821, Faraday had married Sarah Bernard – the daughter of an elder in the Sandemanian church – and she had moved in with him to his Royal Institution lodgings with the Managers' permission. They were to live there until Faraday's death in 1867. In 1833, Faraday was appointed the first Fullerian Professor of Chemistry at the Royal Institution, increasing his income by another £100 a year. For the remainder of his career, the Royal Institution was to be Faraday's base. It provided him with all the resources he needed to pursue his studies. It not only gave him a home and a living; he also had the laboratory at his disposal, as well as the services of the Institution's laboratory assistants and instrument-makers. The Institution paid between £100 and £200 annually for experimental apparatus and other materials for the laboratory. From 1827, Faraday had his own personal assistant in the form of Sergeant Anderson. If the Royal Institution was vital for Faraday's career, however, Faraday was also vital for the Royal Institution's reputation. For much of the second quarter of the 19th century, while Faraday was at the height of his reputation and scientific activity, he and the Royal Institution were practically indistinguishable.

The Royal Institution's managers were well aware of what they had in Faraday. As the Duke of Somerset, then President of the Royal Institution, wrote to the mathematician Charles Babbage in the mid-1830s: 'The story of Faraday is sure to make a great noise. There is something romantic and quite affecting in such a conjunction of Poverty and Passion for Science, and with this and his brilliant success he comes out as the Hero of Chemistry.' As Somerset well knew, if Faraday's story was to make a 'great noise', then so would the Royal Institution's. Faraday, in turn, was extremely proud of the Institution and jealous of its reputation. When William Upcott, assistant librarian at the rival London Institution, dared to cast doubt on its preeminence, Faraday lashed out waspishly. 'I am amused and a little offended by Upcots [*sic*] hypocrisy. He knows well enough that to the world an hours existence of our Institution is worth a years of the London and that though it were destroyed still the remembrance of it would live for years to come in places where the one he lives at has never been heard off [*sic*]. ... I think I could make the man wince if I were inclined and yet all in mere chat over a cup of tea.'

Just what the Royal Institution's and Faraday's reputation was, becomes clear in a particularly ironic passage in the English writer George Eliot's novel, *The Mill on the Floss* (1860). Eliot casts an amused and sardonic eye over the pretensions of mid-Victorian fashionable society and its foibles. 'In writing the history of unfashionable families', she wrote, 'one is apt to fall into a tone of emphasis which is very far from being the tone of good society, where principles and beliefs are not only of an extremely moderate kind, but are always presupposed, no subjects being eligible but such as can be touched with a light and graceful irony. But then, good society has its claret and its velvet carpets, its dinner-engagements six weeks deep, its opera and its fairy ballrooms; rides off its *ennui* on thoroughbred horses, lounges at the club, has to keep clear of crinoline vortices, gets its science done by Faraday, and its religion by the superior clergy who are to be met in the best houses: how should it have time or need for belief and emphasis?' This was the audience that Faraday catered for at the Royal Institution. His success there depended on his being able to give fashionable London what it

Illustration 7: The impressive façade of the Royal Institution on Albemarle Street. Michael Faraday spent his entire professional career as a natural philosopher working in this building.

wanted by way of science. What fashionable London wanted, by and large, was a polished performance and a carefully sanitised view of nature that was, as the polymathic William Whewell, Master of Trinity College, Cambridge, put it, 'devoid of extraneous machinery'.

There was, of course, a fair amount of 'extraneous machinery' that went on behind the scenes at the Royal Institution. Faraday's talent as a public performer lay, in part at least, in his ability to disguise the work that went into making his performances perfect. Everything that Faraday did in public, in front of his genteel audience in the Royal Institution's lecture theatre, was carefully prepared beforehand. Apparatus was tried and tested, experiments were diligently rehearsed, so that by the time Faraday presented them in the lecture theatre it looked, ironically enough, as if Faraday himself was doing nothing. One of the secrets of Faraday's success – as Whewell hints – lay in his ability to make it clear that nature, rather than Faraday himself, was calling the shots. This is why the Royal Institution was so vital for Faraday's career. It had the resources that made possible the kind of meticulous preparation that was essential in order to produce the seamless performances for which Faraday became famous. His performances started life in the basement laboratory. By the time they were carried upstairs to the public arena they had been carefully honed and polished, so that very little evidence remained of the teams of assistants, the forges and furnaces, the foul-smelling galvanic batteries – and the sheer repetitive labour – that had been needed to make them possible.

Laboratory preparation in the basement was not Faraday's only back-room activity, either. Increasingly during the 1820s, Faraday took charge of the Royal Institution's day-to-day routine activities. As the Superintendent of the House (as well as Director of the Laboratory) he was responsible for the upkeep of the building and supervision of its servants. It was his job to hire and fire. It fell to him to make sure essential repairs were carried out, such as in early 1830 when he reported to the Managers that the success of the Friday Evening Discourses was in danger of bringing the house down in a far too literal sense. It seemed that the size of the audience was causing the Library floor to sink under its weight, and an iron pillar had to be erected in the Newspaper Room below to prevent the subsidence. The Royal Institution's hectic lecture schedule had to be organised – and Faraday was the man to do it. Guest speakers had to be invited to deliver those Friday Evening Discourses that Faraday did not give himself. By the 1830s, Faraday had fashioned himself not only into a superlative scientific lecturer, but into the Royal Institution's indispensable man-about-the-house as well. Neither the institution nor the individual could survive without the other.

· CHAPTER 13 ·

SELF-FASHIONING

From the beginning of his career as Sir Humphry Davy's philosophical apprentice – indeed, if not sooner – Faraday was keenly aware that making a success of a scientific vocation would need more than raw talent and aptitude. This was one of the first things he had learned from John Tatum at the City Philosophical Society and from his avid reading of Isaac Watts's *The Improvement of the Mind*. These sources taught him that, above all else, he needed to cultivate self-discipline in order to succeed. This was also something that Faraday learned from Sir Humphry Davy. It is probably what he had in mind with his catty remark regarding Davy, 'that the greatest of all his [Faraday's] great advantages was that he had a model to teach him what he should avoid'. What Faraday meant by this was that Davy's example had shown him that he could not hope to succeed by simply trying to become one of his betters, as Davy had tried to do. If he wanted to survive in the world of gentlemanly natural philosophy, he had to find a different strategy from the one of assimilation that Davy had followed. This was a message that would certainly have been reinforced during his Continental tour by Lady Jane Davy, with her constant reminders of just what Faraday's proper place should be. Faraday, accordingly, invested a great deal of effort in developing a persona for himself that allowed him to move successfully in the world of gentlemanly science, without quite becoming a gentleman of science himself.

Faraday's early correspondence with his close friend Benjamin Abbott shows just how important he thought it was that a scientific lecturer should behave in a particular fashion, and how carefully his surroundings should be organised to achieve the best effect. From his early days at the Royal Institution he paid particular attention to how others performed – 'how the audience were affected and by what their pleasure & their censure was drawn forth'. He was well aware of the importance of conforming to his future audience's preconceptions of the proper style of natural philosophical presentation. 'A Lecturer may consider his audience as being Polite or Vulgar', he lectured his correspondent, 'Learned or unlearned (with respect to the subject) Listeners or Gazers – Polite Company expect to be entertained not only by the subject of the Lecture but by the manner of the Lecturer, they look for respect, for language consonant with their dignity and ideas on a levell [*sic*] with their own.

The vulgar that is to say in general those who will take the trouble of thinking and the bees of business wish for something that they can comprehend. This may be deep and elaborate for the Learned but for those who are as yet Tyros and unacquainted with the subject must be simple and plain. Lastly Listeners expect reason and sense whilst Gazers only require a succession of words.'

Experiments needed to be carefully planned and adapted to the purposes of the lecture, even down to their arrangement on the lecture table: 'Every particular part illustrative of the Lecture should be in view no one thing should hide another from the audience nor should anything stand in the way of or obstruct the Lecturer – they should be so placed as to produce a kind of uniformity in appearance.' They needed to be presented to the audience in a particular way – nothing too showy or ostentatious, and no unnecessary drawing attention to the lecturer's own mastery of the experiment. Comments about the actual process of experimentation itself were necessary only if something went wrong, and even then should be kept to a minimum. The lecturer needed to know how to move his own body – 'he must by all means appear as a body distinct and separate from the things around him and must have some motion apart from that which they possess'. This was clearly a topic that fascinated Faraday at the beginning of his career as much as it became vital to him in later life. His enthusiastic scribblings to Benjamin Abbott on the pros and cons of successful self- and scientific presentation continued through four long letters in June of 1813. All the indications are that he learned the lessons they suggested well.

Faraday's favourite bedside reading of the early 1810s, Isaac Watts's *The Improvement of the Mind*, singled out lecturing as the most pleasant of the five ways of learning: observation, reading, instruction by lectures, conversation and meditation. 'There is something more sprightly, more delightful and entertaining in the living discourse of a wise, learned, and well-qualified teacher', Faraday would have read, 'than there is in the silent and sedentary practice of reading. The very turn of voice, the good pronunciation, and the polite and alluring manner which some teachers have attained, will engage the attention, keep the soul fixed, and convey and insinuate into the mind the ideas of things in a more lively and forcible way, than the mere reading of books in the silence and retirement of the closet.' The correspondence with Abbott shows that Faraday took Watts's recommendations to heart. When opportunity arose, Faraday took advantage of the chance to provide for himself the 'turn of voice, the good pronunciation, and the polite and alluring manner' that Watts thought so important, by taking lessons with Benjamin Smart, one of

London's foremost elocution teachers. As noted earlier, this was not a trivial undertaking for an impecunious young man. That Faraday was prepared to invest half a guinea per lesson on the process is an indication of the importance he placed on acquiring the skills that Smart could teach him.

Smart's teaching took his pupils through three stages of learning on the way to elocutionary success. They were taught mechanical reading to show them how to pronounce words 'justly, completely, and in smooth unbroken series between the written stops when they are joined into sentences'. They were taught significant reading to make 'the construction and meaning of every sentence plain by appropriate *tunes* (or inflections) of the voice'. Finally, they achieved the dizzy heights of impassioned reading, or '*distinct, significant, impressive* Speaking', which was the final goal of elocution. This involved more than simply using the voice to best advantage: 'the whole looks, the gesture, the whole deportment of the speaker, lend assistance; and it is the union of all these that constitutes *expression.*' Smart gave his pupils detailed instructions on how they should stand and hold their bodies to maximum effect. He taught them how to use their eyes and hands in order to give the impression of '*laying*, as it were, his facts or truths *before* his auditors', and urged constant practice to convert these conventional gestures into natural bodily motions. When Faraday first started lecturing at the Royal Institution he invited Smart and others such as Edward Magrath, an old acquaintance from the City Philosophical Society, to come along and judge his performances. They would even hold up placards saying 'too fast' or 'too slow' as guides to Faraday's delivery.

Faraday's aim in all of this was to transform himself into the gentleman's natural philosopher. There was a subtle distinction between that goal and the one that Humphry Davy had aspired to – to become a gentleman of science himself. Faraday fashioned himself as nature's spokesman for the leisured classes, rather than trying to become one of them himself. In 1836, the scandalous *Fraser's Magazine* published a profile of Faraday predicting imminent elevation to a knighthood: 'he is now what Davy was when he first saw Davy – in all but *money*', they proclaimed – joking that '*Far-a-Day*' was easily translated as '*Near-a-Knight*'. They were wrong, of course, showing only how little the Tory rag understood their hero – despite Faraday's own Tory leanings. Faraday had no intention of aping his betters as he thought Davy had done. Throughout his career, Faraday kept both his audience and his gentlemanly scientific contemporaries at a careful arm's length. He rarely attended social gatherings outside the Royal Institution, such as the gregarious Charles Babbage's glittering scientific soirées. He even used his Sandemanianism (a religious affiliation that completely baffled his largely moderate Anglican audience) as a foil to keep outsiders at bay. This was quite deliberate. His espousal of such an outré sect was in part, at least, a way of distancing himself from the society in which he was obliged to practise his scientific calling.

· CHAPTER 14 ·

PERFORMING SCIENCE

Faraday recognised clearly that a large proportion, at least, of his Royal Institution audience would fall into the category of Gazers, rather than Listeners. By the late 1820s, when Faraday was fully established there as Director of the Laboratory, the Royal Institution had already made its reputation as the metropolis's premier venue for purveying natural philosophy to the fashion-conscious. The relationship between science and showmanship was well established by Faraday's time. At the Royal Institution itself, his predecessor Sir Humphry Davy had made the Institution's reputation – and his own – through flamboyant and spectacular displays of nature's powers. With his combination of rhetorical flair and skilful experimentation he had transformed the humble apothecary's apprentice from Penzance into the epitome of the fashionable philosopher. By the 1830s, science at the Royal Institution was firmly established at the higher end of the range of entertainments available to London's leisured classes. For much of the Royal Institution's clientele, going to a scientific lecture there was much like going to the theatre, or a society soirée. They went there to learn something, certainly. But they also went there to see and be seen. They went there to enjoy themselves and to be thrilled by new things. They expected to be moved by their experiences and they would judge Faraday, like any other stage performer, on the basis of his success in providing them with the kinds of feelings and sensations they expected to receive.

There was a range of different kinds of scientific performances available at the Royal Institution as well, tailored to different kinds of audiences. As we saw earlier, the Institution's lecturing schedule was reorganised at Faraday's suggestion in the late 1820s so as to conform more closely to the rhythms of the London Season. The exception was a series of morning lectures for medical students (a lucrative market in early 19th-century London) that ran from October each year. Lectures for ticket-holding members and their guests took place in the afternoons. Another of Faraday's innovations was the Christmas series of Juvenile lectures (a tradition that continues today) aimed at children. A famous painting by Alexander Blaikley shows Faraday performing at the Christmas lecture series for the 1855–6 season. The painting seems to depict at least as many adults as children in the audience – and as many women as men. Seated prominently in the front row is Albert, the Prince Consort, flanked by the future Edward VII and his younger brother. Most popular of all were the Friday Evening Discourses. These affairs were extremely prestigious and tickets for them were keenly fought for. Faraday was extremely careful to make sure that these precious pieces of paper did not fall into the wrong hands. On one occasion at least, he begged a ticket holder to burn his ticket if he could not use it himself.

Topics for the Friday Evening Discourses were carefully selected. Faraday tartly advised one hopeful aspirant for a lecturing slot that 'New points in philosophy – or new modes of experimental illustration – or new applications to useful purposes are what is wished for, after these follow new matters in taste or literature. But mere matter of opinion which can be settled only by reference to taste and not by reference to natural facts is ... inadmissible at our Lecture table on the Friday Evenings.' Anything that might be considered offensive to the audience's sensibilities was rigorously excluded. Early topics ranged from Marc Brunel's block machinery recently installed at the Portsmouth Naval Dockyards (which de-skilled a generation of shipbuilders) to the latest developments in electricity. In 1832, for example, Faraday gave three discourses on his own recent work on electricity and magnetism. Faraday did not restrict himself solely to his own specialities in his

Illustration 8: Michael Faraday performing at one of the Royal Institution's Christmas lectures. Sitting in the front row precisely in front of him are Prince Albert, the Prince Consort, and the Prince of Wales.

discourses, however. In 1837, for example, he gave a talk on 'Early Arts: the Bow and Arrow' which was, as one female guest recalled, 'most interesting and amusing, and of course well delivered. Mr. F. shot or rather blew several small arrows through tubes – and with good aim – at a band box with a centre mark.' The Discourses were extremely popular, with the numbers attending frequently in the hundreds. The Discourse at which Faraday so impressed his female guest with his marksmanship was attended by 583 admirers, leading her to complain: 'place full, but the heat and draught dreadful.'

Faraday's apparent facility with a blowpipe in 1837 hints, if a hint is needed, that his performances were not entirely off-the-cuff. The same female guest, after attending another discourse two years later, gushed that Faraday was 'the *beau idéal* of a popular lecturer'. Commentators often remarked on how unforced his lecturing style appeared – 'his manner was so natural that the thought of any art in his lecturing never occurred to anyone'. As one of his early biographers made clear, however, for 'his Friday discourses, and for his other set lectures in the theatre, he always made ample preparation beforehand. His matter was always over-abundant, and, if his experiments were always successful, this was not solely attributable to his exceeding skill of hand. For, unrivalled as he was as a manipulator, in the cases in which he attempted to show complicated or difficult experiments, that which was to be shown was always well rehearsed beforehand in the laboratory.' Faraday had clearly done a good job of following Benjamin Smart's advice to his pupils that they should work at turning the stylised and conventional gestures of oratorical art into natural bodily movements. He had fashioned himself into the ideal type of a natural philosopher for his genteel audience. He was clearly a success for the Royal Institution as well, turning the precarious financial position of the early 1820s into a healthy surplus by the 1830s.

Faraday's performances certainly made an impression on those who saw them. As one susceptible audience member later recalled the experience, 'it was an irresistible eloquence, which compelled attention and insisted upon sympathy ... There was a gleaming in his eyes which no painter could copy, and which no poet could describe. Their radiance seemed to send a strange light into the very heart of his congregation, and when he spoke, it was felt that the stir of his voice and the fervor of his words could belong only to the owner of those kindling eyes. His thought was rapid, and made itself a way in new phrases – if it found none ready made – as the mountaineer cuts steps in the most hazardous ascent with his own axe. His enthusiasm sometimes carried him to the point of ecstasy when he expiated on the beauties of Nature, and when he lifted the veil from her deep mysteries. His body then took motion from his mind; his hair streamed out from his head; his hands were full of nervous action; his light, lithe body seemed to quiver with its eager life. His audience took fire with him, and every face was flushed.' This was a carefully cultivated image. Faraday wanted his audience to see him in a certain light (though he may not have expected them to have entered into things quite as enthusiastically as Lady Pollock seems to have done) because he wanted to convey to them a particular image of the kind of person a natural philosopher should be, and a particular image, therefore, of what science itself should be.

Faraday, the skilled experimental manipulator of the laboratory, was replaced in the lecture theatre by Faraday, the Romantic genius. His lectures were designed to appeal to the imaginations of the Gazers in his audience at least as much as to the reason of the Listeners. John Tyndall, one of Faraday's successors at the Royal Institution and one of his first biographers, commented that Faraday's 'Friday evening discourses were sometimes difficult to follow. But he exercised a magic on his hearers which often sent them away persuaded that they knew all about a subject of which they knew but a little.' Cornelia Crosse (the insect-making Andrew Crosse's wife) made a similar point in slightly more flattering terms: 'No attentive listener ever came away from one of Faraday's lectures without having the limits of his spiritual vision enlarged, or without feeling that his imagination had been stimulated to something beyond the mere exposition of physical facts.' The image of himself that Faraday conveyed on the Royal Institution's stage was highly idealised. So was the image of science that he communicated. The science that Faraday presented was designed to appeal to the sentiment of his audience. It seemed divorced from the hard work that had to go on behind the scenes to make it appear so effortless on stage.

· CHAPTER 15 ·

DRAWING THE BOUNDARIES OF SCIENCE

One face that Faraday certainly presented to his Royal Institution audience was the face of an expert. This was an increasing trend during the first half of the 19th century. The new generation of gentlemanly men of science that came to the fore during the 1820s regarded themselves as disciplined experts in their various scientific specialisms. Specialisation in science was a new trend too – typified by the growth in specialist scientific societies such as the Geological Society (founded in 1808), the Astronomical Society (1820) and the Chemical Society (1844). These gentlemanly specialists – men like the astronomer John Herschel, the geologist Charles Lyell, or the mathematician Charles Babbage – saw their science as a vocation. They were entitled to speak on behalf of nature because they were trained experts who had spent years doggedly acquiring the skills that were needed to do the job properly. They were contemptuous of the previous generation's dilettantism and amateurishness (as they saw it). They were equally contemptuous of the ragbag of radicals and charlatans (again, as they saw them) who offered different visions of what science was about, whom it was for and how it should be ordered. In 1833, William Whewell invented the word 'scientist' to describe a new kind of scientific worker – someone who worked under the direction of philosophers like him, but could not, themselves, aspire to the title of philosopher. He had the membership of the recently established British Association for the Advancement of Science in mind. The word took a very long time to catch on. Someone like Faraday certainly never had any truck with it. Faraday had no doubt in his mind that he was a philosopher.

Faraday, though not himself a gentleman of science, shared many of their views concerning the proper practice of science and the kind of individual who could be trusted to carry out that task. Science and the public needed protecting from false prophets and needed to be taught whom they should and should not trust in matters requiring scientific judgement. It was a matter of drawing boundaries around science. This was another reason why Faraday was so careful about policing who could and could not be allowed to deliver Friday Evening Discourses at the Royal Institution. Faraday's response to the craze for mesmerism that swept England during the 1830s and 40s was symptomatic. Mesmerism had been imported from France towards the end of the 18th century – one among many revolutionary philosophical enthusiasms of the period. Mesmerists claimed to be able to take over their subjects' minds and control their bodies. Proponents hailed it as the basis for a new, materialist science of the mind. Faraday, on the other hand, condemned it as an example of charlatanry working away at the minds of the unlearned. The proper antidote to mesmerism was a good dose of mental education, which would teach people to discriminate between true and false phenomena and – more importantly, perhaps – between true and false reporters of phenomena.

Faraday had similar things to say about the vogue for table-turning that swept Britain in the early 1850s. Table-turning, on the face of it, seemed a simple enough phenomenon. A group of hopeful table-turners would meet for a *séance* and sit around a circular table. They would each place their hands, palm downwards, on the table and wait. Under their combined mental influence – if the experiment worked, that is – the table would spontaneously start to move, apparently of its own accord. Faraday devised an ingenious experimental arrangement to show that, in reality, the hands rather than the table were doing the pushing. Two thin wooden boards were placed on top of each other on the table, separated by a number of glass rods lying between them. A straw, attached to the boards by a couple of pins, stuck out between them as an indicator. If, during a table-turning experiment, the straw was deflected in one direction, it meant that the turning force came from the table. If it was deflected the other way, the turning force came from the hands placed on it. Faraday's argument was that table-turning was another example of the triumph of hope over experience. Participants wanted the experiment to work, so they unconsciously made sure that it did. It was all too easy, Faraday argued, for even the best-intentioned but untrained minds to fool themselves over what the phenomena (and the true causes of the phenomena) really were. Table-turning, like mesmerism, was simply wishful thinking. That was why it was so important that they be taught to trust experts like Faraday himself, whose minds had been trained to distinguish between fact and wish-fulfilling fiction.

In debates such as these, Faraday was acting the part of policeman, patrolling the boundaries of science. He acted in a similar way over the controversy surrounding Andrew Crosse and his electrical creation of life. It was not just the occasional 'infidel lecturer' who alleged that Faraday had succeeded in producing electrical insects like Crosse's. There was even a report in *The Times* to that effect, to which Faraday penned an angry denial. He also carried out some experiments of his own, on some electric eels borrowed from the

Adelaide Gallery, to demonstrate what in his view were the acceptable boundaries of experiments on the principle of life itself. Experiments like these were 'upon the threshold of what we may, without presumption, believe man is permitted to know of this matter', he said. They were designed to show that the eels' electricity was the same as electricity from other, inorganic sources, and to investigate the relationship between the electricity they produced and the exhaustion of their nervous force. But, Faraday argued – against the electrical radicals who had pounced on Crosse's experiments with such glee – the experiments showed that the nervous force was not itself the principle of life. It was just a force like any other, and therefore a legitimate subject of experimental investigation. Investigating the mysteries of life itself, however, was for Faraday entirely beyond the scientific pale.

In all these examples, Faraday was proffering himself as a model of the ideal man of science, and the Royal Institution as a model of the ideal scientific institution. There was nothing particularly disingenuous or self-serving about this. There is certainly no reason to doubt that Faraday was perfectly sincere in his views concerning the proper practice of science. The early 19th century was a period when the definition of what counted as science, scientists and scientific institutions was very much up for grabs. Our modern understanding of what counts as good science is still to a large degree based on the outcome of these early 19th-century debates. For Faraday, science was about expertise. Men of science were the product of hard work, but they were also the product of a particular cast of mind. Faraday may have acknowledged the important role played in his own early career by Humphry Davy's tutelage, but he just as clearly felt that Michael Faraday himself – the kind of individual he was – played a crucial role in the equation as well. This is certainly how his public persona was constructed. Faraday's science was also about performance, however. Not just science, but scientific expertise had to be communicated to an audience. That is why, in Faraday's view, places like the Royal Institution and their audiences mattered. If nothing else, the Royal Institution and its audience provided Faraday with a space for experimentation.

PART V: CULTURES OF DISPLAY

· CHAPTER 16 ·

ELECTRICAL UNDERWORLD

By the end of the 1830s, Faraday was certainly one of the leading figures of fashionable London science. The well-heeled flocked to his lectures and the Royal Institution's Friday Evening Discourses. There were still places where Faraday's writ did not run, however. He defined electrical science for genteel society; he did not necessarily define it for everyone else. Beneath the social and intellectual stratum inhabited by natural philosophers like Faraday (and often beneath their notice as well), a city such as London sustained a significant number of other, more humble, practitioners. Just as Faraday's science at the Royal Institution could not take place without the participation of those who laboured behind the scenes, nor could London science take place without the networks of chemists, instrument-makers and other craftsmen, as well as occasional lecturers, that sustained it too. By and large, these were not the kind of people who were likely to gain easy access through the portals of the Royal Institution, as either performers or members of the audience, though they might be able to get in by the servants' entrance instead. This was the London scientific scene that the metropolis's practical electricians inhabited. They moved between instrument-makers' shops, craftsmen's workshops, exhibitions and lecture theatres far less exalted than that at the Royal Institution. It was a very different world from the world of polite science – and its inhabitants consequently often had very different views on what electrical science was all about.

Early 19th-century instrument-makers produced a very wide array of electrical apparatus. They produced different kinds of batteries, and electrical coils of various sorts. As mentioned earlier, during the 1830s and 40s they often directed their products at medical practitioners as devices for delivering medical electricity. Following the development of electromagnetic engines during the 1830s and telegraphs during the 1840s, they often advertised working models of such machines, just as they still also advertised working models of steam-engines or locomotives. They manufactured a whole array of apparatus – like Ampère's cylinder, Barlow's wheel and Marsh's pendulum – designed to show off the relationship of electricity and magnetism in spectacular fashion. They sold galvanometers and other devices to assess electrical output. Much of what they made and sold would, however, have looked perfectly familiar to electrical instrument-makers of the previous generation or earlier. They still sold old-style electrical machines, consisting of plates or cylinders of glass mounted in a frame so that they could be turned by a handle, producing a supply of static electricity. They sold batteries of Leyden jars for storing electricity. They also made a whole range of philosophical toys, as they were often called, like the electrical spider, the screaming head – a wooden head whose hair would stand on end when attached to an electrical machine – or the thunderhouse, which demonstrated the effects of a lightning strike on a building.

Practical electricians such as these often regarded the cosmos as being constituted from electrical machines just like the ones they made. That electricity played a vital role in the operations of the universe was something they took for granted. Few practical electricians would have had much objection to the basic idea, at least, of T. Simmons Mackintosh's electrical universe. As far as they were concerned, the universe was a huge electrical machine and operated by just those principles that underlay the workings of such a contraption. In this respect, as in many others, their view was very different from Faraday's. For Faraday, much as he regarded them as absolutely central to the process of scientific inquiry – and much as he excelled at performing them both in private and in public –

experiments were a means to an end. They were a way of finding out about something else. For many practical electricians, on the other hand, experiments simply were the end. Natural philosophy was about finding new and ingenious ways of showing off the powers of nature, and experiments were how one did it. Experiments and machines were what made up the universe. Spectacular displays and demonstrations were what electricity was all about.

Practical electricians' view of natural philosophical practice and its practitioners could be very different from Faraday's as well. The gap between natural philosophers and instrument-makers had widened considerably during the first half of the 19th century. During the 18th century, instrument-making could, on occasion at least, be a route to philosophical fame and respectability. Some instrument-makers even reached the heights of Fellowship of the Royal Society. By the 1830s or 40s this was certainly no longer the case. Increasingly, instrument-makers came to be relegated to the status of support staff rather than potential producers of new knowledge in their own right. Making new knowledge – the business of discovery – was strictly the purview of the highly trained and specialist élite. Unsurprisingly, practical electricians themselves did not usually share this view. As far as they were concerned, they were legitimate knowledge-producers too. Finding new electrical phenomena and contriving novel ways of putting them on display were significant scientific discoveries for them. Faced with increasing marginalisation within respectable scientific circles, practical electricians looked around for new ways of organising themselves, and new institutions within which they could operate. They developed their own distinctive vision of what electricity was about and how it should be practised, which was often at odds with that of the scientific élite. One of the main players in this respect was the electrician William Sturgeon.

· CHAPTER 17 ·

WILLIAM STURGEON AND THE LONDON ELECTRICAL SOCIETY

William Sturgeon was born in 1783 in the village of Whittington in Lancashire. His father, John Sturgeon, was a shoemaker from Dumfries in south-west Scotland and his mother, Betsy Adcock, was the daughter of a local shopkeeper. Sturgeon was apprenticed to another local shoemaker at the age of thirteen, where, according to one biographer, he was 'doomed to be even more cruelly treated than at home'. In 1802, following his qualification as a journeyman shoemaker, Sturgeon enlisted in the Westmoreland Militia. Two years later he joined the 2nd Battalion of the Royal Artillery. In 1804 he married Mary Hutton and they had three children, none of whom survived infancy. After his first wife's death, he married Mary Bromley in 1829. Again, their only child died in infancy and they eventually adopted Sturgeon's niece, following her own mother's death. Sturgeon left the army in 1820 and briefly set himself up as a shoemaker in Lancashire before moving to Woolwich near London and setting up shop at 1 Artillery Place. Sturgeon developed his interest in natural philosophy, and electricity in particular, during his time in the army. Reputedly, he first become fascinated by electricity after witnessing 'a terrific thunderstorm' during a tour of duty to Newfoundland in Canada. He had taught himself mathematics, optics and other branches of science using books borrowed from a sympathetic sergeant. After settling in Woolwich he started to produce his own philosophical apparatus and to supplement his earnings as a shoemaker by making and selling instruments and by occasional scientific lecturing.

Sturgeon's activities as an instrument-maker soon started bringing him to the attention of local natural philosophers around Woolwich. Peter Barlow, the Professor of Mathematics at the nearby Royal Military Academy, was a particular patron, and with his support Sturgeon eventually acquired a post lecturing on natural philosophy at the East India Company's Military Academy at nearby Addiscombe. In 1824, Sturgeon submitted for consideration by the Royal Society of Arts a set of table-top electromagnetic apparatus of his own invention. He was awarded a silver medal and 30 guineas by the Society for his contrivance. The point of Sturgeon's apparatus, as he made clear in his account of it to the Society of Arts, was to find ways of displaying electromagnetic powers as economically as possible by using powerful magnets to compensate for the use of relatively small and cheap batteries. As Sturgeon noted, 'with small magnets the experiments can never be made on a large scale, although the Galvanic force be ever so powerful; and as minute and delicate experiments are not calculated for sufficiently conspicuous illustration in public lectures, I considered that an apparatus for exhibiting the experiments on a large scale, and with easy management, would not only be well-adapted to the lecture room, but absolutely valuable to the advancement of the science'. One item of Sturgeon's table-top apparatus would turn out to be particularly significant: a small, horseshoe-shaped bar of soft iron with a coil of insulated copper wire loosely wrapped around it. This was the first electromagnet, and Sturgeon's invention was to play a critical role in the development of electrical technologies like the electromagnetic engine and the telegraph during the coming decades.

Sturgeon placed particular emphasis on the fact that his apparatus helped to make the detailed operation of electromagnetism visible to his audiences. 'It can be no small gratification to those who are in the habit of giving public lectures', he suggested, 'to be enabled to exhibit this experiment to the satisfaction of a large audience; for as the lecturer can now have his rotating magnet of any size he pleases, and likewise of any figure, this interesting experiment may be viewed from the remotest part of the room'. Being able to view the technical details of the apparatus in action mattered because the apparatus literally represented nature. An experiment in which Sturgeon made a globe constructed from silver and platinum wires rotate around a central magnet when it was heated (since heating the junctions of the silver and platinum wires produced electricity by thermo-electricity) is a case in point. Sturgeon argued that the experiment showed how the Earth rotated on its axis. He detailed the ways in which different terrestrial features could be shown to conform to his model. He showed how the prevalence of thunderstorms in the tropics, for example, was explained by his model since, as the Sun's heat in those regions of the globe was greater than it was nearer the poles, so more thermo-electricity would be generated and hence more atmospheric electricity, thus causing the thunderstorms.

The aim of experimentation, in Sturgeon's view, was to reproduce the workings of the natural economy. The world, after all, operated not only on the same principles, but using the same mechanisms as his electromagnetic apparatus. 'That electrical currents are continually flowing in the earth', he insisted, 'must appear obvious to every one conversant with voltaic electricity. The materials which form our batteries, and display electric streams at our pleasure, have all been brought from this exhaustless source. Nature's laboratory is well stored with apparatus of this kind, aptly fitted for incessant action, and the production of immense electrical tides; and the insignificancy of our puny contrivances to mimic nature's operations, must be amply apparent when compared with the magnificent apparatus of the earth.' This view of experiment and electricity was a far cry from Faraday's perspective. Electricity, for Sturgeon, was a hands-on science. Making the details of the apparatus visible to an audience mattered because that was tantamount to making the details of the operations of nature visible as well. Faraday, on the other hand, saw experiment as being aimed at the production of facts rather than spectacular demonstrations. His aim when he performed in front of his Royal Institution audience was to draw their attention away from the apparatus. Sturgeon's aim, in the far more humble settings in which he performed, was precisely to draw his audience's attention towards the apparatus.

Electrical experimentation, according to Sturgeon, was also a fundamentally communal activity. This was the view that underlay his key role in establishing the London Electrical Society. On 16 May 1837, a number of 'gentlemen' held a meeting at the instrument-maker E.M. Clarke's shop (or 'Laboratory of Science' as he called it) on Lowther Arcade, just off the Strand. They were instrument-makers, operative chemists and struggling lecturers rather than members of London's scientific élite. That, as far as Sturgeon was concerned, was rather the point of the exercise. The members would be 'individuals previously unknown in the annals of science, who have, within the last few years, devoted their time as well as pecuniary means to the cultivation of electricity'. The Society would be 'a parent to foster and cherish their investigations; a grand storehouse in which they may repose the rich productions of their labours, and a temple for their kindred spirits' resort'. Sturgeon thought the élite assertion that only the trained specialist could make a real contribution to electricity, or any other science, ridiculous – 'as groundless as it is detrimental to the progress of any particular branch' of science. Participation in the proceedings of the London Electrical Society would be open to all comers. Anybody who had eyes to see and witness electrical phenomena could make a valuable contribution.

Its promoters hoped that the Society would prove a breeding ground for a new generation of experimenters. 'At the present time', noted Henry Noad in his *Lectures on Electricity*, 'particular advantages are laid open to the tyro. An electrical society has been recently established in London, and if the future may be prognosticated from the present, its success will be triumphant; rarely a month passes, without some new and important fact being announced, or some new apparatus being exhibited; conversation is unrestricted; and here the beginner may, from the experience of the more advanced enquirer, get his difficulties removed ... the information derived from books, though it may do well for the closet, will generally be far from satisfactory in the laboratory; here more detailed instruction, particularly in manipulation is required, and it is by actually witnessing the various operations performed that the necessary information can be acquired; hence the great advantage of a society, in which there is a community of taste and feeling, and in which knowledge is unrestrictedly communicated.' Noad, like Sturgeon, regarded the Society as a self-help group in which the members would work together for their mutual improvement – a not uncommon aspiration for men like them, inhabiting the precarious social ground between the lower and the middle classes.

The Society held regular gatherings at which its members would perform experiments, demonstrate new apparatus and novel electrical devices or deliver lectures. E.M. Clarke, at whose 'Laboratory of Science' the early meetings were held, showed off his latest magneto-electric machine, demonstrating how its effect 'on the nervous and muscular system is such, that no person out of the hundreds who have tried it could possibly endure the intense

Illustration 9: The frontispiece of Henry Noad's *Lectures on Electricity* (1844). At centre stage is an Armstrong hydro-electric machine. Various items of electricians' apparatus are scattered around the room.

agony it is capable of producing', and how it gave off 'large and brilliant sparks, sufficiently so that a person can read small print by the light it produces'. The American inventor Thomas Davenport's electro-magnetic engine was put on show when its promoters visited London. William Sturgeon gave regular accounts of his 'Experimental and Theoretical Researches in Electricity &c.' and William Leithead delivered his views on the relationship between electricity and disease. At one of the Society's earliest meetings, a letter containing the first detailed account of Society member Andrew Crosse's production of insect life by electricity was read to the gathering. Another member, James Prescott Joule, the son of a Manchester brewer, enthusiastically corresponded with the Society, sending them accounts of his efforts to produce the most efficient electro-magnetic engines possible. The Society's activities rode rough-shod over the sorts of nice distinctions between science and invention, or between different scientific disciplines, that respectable natural philosophers like Faraday increasingly insisted upon.

In practice, however, the Society fell very far short of its utopian aspirations towards electricity as an egalitarian science and experiment as a communal activity. The Society was soon riven by internal rivalries, with the thin-skinned and combative Sturgeon at the centre of things. Towards the end of the 1830s, Sturgeon left the Society he had been instrumental in establishing in a huff, following a major row with Charles Vincent Walker, the new honorary Secretary. Walker felt that Sturgeon was unfairly trying to take undue credit for some observations made during a day's communal experimenting at the house of one of the Society's members, Thomas Mason. In Walker's eyes, Sturgeon was asserting his own individual intellectual property rights over discoveries that ought to belong to the Society communally. Sturgeon, in high dudgeon, departed – in fact he departed from London entirely and moved to Manchester in 1840, where he became the Superintendent of the Royal Victoria Gallery for Practical Science. A few years later, following a hiatus in its activities, the London Electrical Society folded, with Walker apologising profusely to the members for the financial mismanagement (mainly to do with hugely over-optimistic print-runs of the Society's *Proceedings and Transactions*) that had led to its downfall. The number of the Society's active members had remained small throughout its existence, and with their leader's departure – 'the head of the second rate philosophers of London', as the American natural philosopher Joseph Henry, in a truly barbed and double-edged compliment, described Sturgeon – they proved unable to keep up the momentum.

· CHAPTER 18 ·

ELECTRICITY ON SHOW

For the first few months of its fledgling existence, the London Electrical Society met, as we have seen, at E.M. Clarke's 'Laboratory of Science' on the Lowther Arcade. After that, they moved – across the way, more or less – to hold their meetings in the Lecture Room of the Adelaide Gallery. The Adelaide Gallery was the first of a number of establishments that flourished briefly in London from the 1830s through to the 1850s, and that aimed at bringing 'practical science' to the public in spectacular fashion. These institutions were 'the places in fact which furnish the great mass of the public with demonstrations of science. The other institutions are in a manner exclusive: membership or the introduction of a member is, with a few exceptions, imperative on those who would be present ... But the exhibitions now before us ... are accessible to all who proffer their shilling at the door.' London during the first decades of the 19th century had a flourishing culture of exhibitions. Those looking for entertainment mixed with (at least a little) edification could choose between panoramas, dioramas, waxwork anatomical shows, magic lantern extravaganzas – and even balloon ascents. Natural philosophy and the mechanical arts were part and parcel of this culture of display. Scientific lectures jostled with gothic melodramas for the attention of the theatre-going public. Working models of the latest industrial machinery or experimental apparatus rubbed shoulders with collections of exotic curios and historical memorabilia of all sorts.

The Adelaide Gallery – or the National Gallery of Practical Science, Blending Instruction with Amusement, to give it its full title – was the brainchild of Jacob Perkins, an American inventor and entrepreneur who had been in London since 1819, trying to interest the Bank of England in his machine to engrave forgery-proof money and, when that failed, the Army in his patented Steam Gun. The Adelaide Gallery, when it first opened its doors in 1832, was originally intended as another project to puff Perkins's own productions. It soon acquired a more ambitious brief, to 'promote ... the adoption of whatever may be found to be comparatively superior, or relatively perfect in the arts, sciences or manufactures [and to display] specimens and models of inventions and other works &c. of interest for public exhibition, free from charge ... thereby gratuitously offering every possible facility for the practical demonstration of discoveries in Natural Philosophy, and for the exhibition of any new application of known principles of mechanical contrivances of general utility.' The Adelaide Gallery, in so many words, was designed to be a showcase for invention and therefore for the individual inventors who put their wares on display there. It was not long before electricity took its place in the list of wonders being exhibited.

The Gallery's main attraction was Jacob Perkins's Steam Gun, which was fired up several times a day to discharge rounds of 70 balls in four seconds at a target placed at the further end of the Gallery's Long Room, which also featured a 70-foot-long canal and pool on which demonstrations of model paddle-driven steam-boats were given. E.M. Clarke had a variety of electrical batteries on show, while William Sturgeon – himself one of the Adelaide Gallery's lecturers – exhibited a ferromagnetic globe to demonstrate the origins of the Earth's magnetism. Joseph Saxton – another American, and the Gallery's in-house instrument-maker – had a large electromagnet on show to illustrate the production of an electric spark by magnetism. As one satisfied customer recorded: 'Clever professors were there, teaching elaborate science in lectures of twenty minutes each. Fearful engines revolved and hissed, and quivered. Mice led gasping sub-aqueous lives in diving-bells. Clockwork steamers ticked round and round a basin perpetually to prove the efficacy of invisible paddle-wheels. There were artful snares laid for giving galvanic shocks to the unwary.' One particular fan was the Duke of Wellington, something of an aficionado of London's exhibition scene. In rooms surrounding the upstairs Gallery, visitors could see the Hydro-oxygen microscope magnify the waters of the Thames an alleged 3,000,000 times. Feeding time for the Adelaide's rare electric eels was another major attraction. During the evenings, performances by the latest stage celebrities, like the Infant Thalia (an early Victorian version of Charlotte Church) and P.T. Barnum's General Tom Thumb, intermixed with lectures and magic lantern shows.

The Adelaide Gallery soon acquired a competitor in the form of the Polytechnic Institution, which opened its doors on Regent Street in the summer of 1838. The Royal Polytechnic Institution, as it was known after being granted a royal charter a year later, was founded by the aviation pioneer Sir George Cayley. He was joined by Charles Payne, who had been the Superintendent of the Adelaide Gallery and took the position of the Polytechnic's manager. The Polytechnic aimed to do exactly what the Adelaide Gallery did, only better. The Hall of Manufactures displayed the tools of various industries. Beneath the Hall was a chemical laboratory where the chemist J.T. Cooper helped hopeful experimenters and patentees to carry out their experiments. Above the Hall was the Lecture Theatre, with seating for 500. Beyond was the Great Hall, 120 feet long and 40 feet high, where the main exhibits were on show. Where the Adelaide had Perkins's Steam Gun as its main attraction, the Polytechnic had a diving bell, in which visitors could descend into the depths of a large tank of water for an additional penny. It also had an oxyhydrogen microscope, billed as 'the largest ever seen', and a huge, steam-driven electrical machine. Later in the 1840s this was eclipsed in turn by a yet more powerful source of electricity. The Armstrong Hydroelectric Machine produced electricity by friction as steam was forced at high pressure through an array of small nozzles. The Polytechnic billed it as 'at least six times more powerful than any other electrical machine ever constructed'.

The Polytechnic soon eclipsed the Adelaide Gallery, and sent it out of business by the mid-1840s. As one nostalgic visitor reminisced towards the end of the century: 'Ah me! The Polytechnic with its diving-bell, the descent in which was so pleasantly productive of immanent head-splitting; its diver, who rapped his helmet playfully with the coppers which had been thrown him; its half-globes, brass pillars, and water-troughs so charged with electricity as nearly to dislocate the arms of those that touched them ... with all these

Illustration 10: The main hall of the Royal Polytechnic Institution. The famous diving bell is visible in the background.

attractions and a hundred more which I have forgotten, no wonder that the Polytechnic cast the old Adelaide Gallery in the shade and that the proprietors of the latter were fain to welcome an entire and sweeping change of programme.' Two visitors from further afield were equally impressed by both the Adelaide and the Polytechnic. According to Jehangeer Nowrojee and Hirjeebhoy Merwanjee, 'if we had seen nothing else in England besides the Adelaide Gallery and the Polytechnic Institution, we should have thought ourselves amply repaid for our voyage from India to England'. By and large, these were the sorts of places where practical electricians and their inventions appeared before the public. Spectacular showmanship was an integral part of their activities. To the public who offered their shillings at the door, it was what electricity was all about.

In the mid-1850s, the instrument-maker E.M. Clarke, who had been a prolific exhibitor of electrical inventions at the Adelaide and Polytechnic, opened his own venture, the Panopticon of Arts and Sciences, on Leicester Square. Electrical exhibits were not confined to specifically scientific exhibitions, either. The Colosseum in Regent's Park, famed for its amazing panorama of London as seen from the top of St Paul's Cathedral, also boasted a 'Department of Natural Magic' under the superintendence of yet another member of the

London Electrical Society, the operative chemist William Leithead. Like the Polytechnic, the 'Department of Natural Magic' also featured the largest electrical machine in the world. During the 1860s, the Royal Polytechnic Institution, by then under the proprietorship of 'Professor' J.H. Pepper, featured a massive induction coil, weighing fifteen hundredweight with a primary wire of 3,770 yards and a secondary wire 150 miles in length, to continue its tradition of gargantuan electrical machinery. Electricity also featured strongly in the range of displays on offer at the greatest show on Earth, the Great Exhibition of 1851 at the Crystal Palace in Hyde Park. Visitors to the Great Exhibition could keep time by the Electric Clock hanging above the Main Transept and see a whole array of electrical machines and devices in action.

The connection between electricity and showmanship continued throughout the century. As late as 1892, the medical magazine *The Lancet* carried a light-hearted little note describing an electrical accident in London: 'It seems that through some unexplained failure of the insulator of an electric-lighting main a metal shop front became charged with wandering electricity and passers-by discovered that by touching it with their hands they could experience the delightful inconvenience of an electric shock. The circumstance that this luxury is in the minds of most people, and particularly of most street-urchins, associated with festival occasions and penny fees no doubt invested the unpriced supply in Walbrook with an added charm.' Scientific exhibitions were a central part of later Victorian culture. Nineteenth-century commentators were certainly aware of the pivotal role that exhibitions played in the century's public life. One pundit remarked that 'the institution existed in the latter half of the nineteenth century, because it was one suited to the requirements of the period'. Exhibitions provided a way of bringing electricity, electricians and their productions inside public culture. They were expressions of late Victorians' confidence in their capacity to transform nature and culture through technology. As we shall see, by the end of the century electrical entrepreneurs like Edison or Westinghouse were competing avidly with each other to produce the most extravagant and spectacular of electrical effects at international exhibitions such as the Columbian Exposition in Chicago in 1893.

· CHAPTER 19 ·

UTILITY

One of the reasons, of course, why electrical entrepreneurs like Edison or Westinghouse at the end of the century competed so fiercely to mount electrical exhibitions was that they knew it was a good way of selling their products. The link between electricity, exhibitionism and utility stretched back to the beginnings of 19th-century electrical exhibition culture, however. Electrical machinery and potential industrial processes of various kinds were regularly exhibited at meetings of the London Electrical Society and at Galleries such as the Adelaide and the Polytechnic. Utility was central to the vision that practical electricians had of the progress of their science. As Henry Noad argued, 'when we consider the almost certainty of its usurping, at no very distant period, the place of steam, as a mechanical agent, and thus being made, in the most extensive manner, subservient to the uses, and under the control of man, little more inducement, will I imagine, be wanting to increase the body of its cultivators, and to assign it to its proper place, among the most important of the physical sciences'. Showmanship was an important element in the business of invention. To win friends and influence people, inventors needed to put their inventions on display, and electrical inventors were no exception. Exhibition, in the public eye, was quite simply part of what an electrician did.

The example of Edward Davy, one of the early British electric telegraph pioneers, is a case in point. Even before patenting his telegraph design, Davy hired private rooms at Exeter Hall in London to put his invention on show before the public. He remarked in a letter to his father that 'You did not expect to have a son turned showman, but I trust I am merely instrumental in promulgating a useful discovery'. Similarly, once Charles Wheatstone and William Fothergill Cooke had secured the patent rights to their eventually triumphant scheme of electric telegraphy, they had to find ways of making their invention public and drawing the attention of potential investors. Exhibition was a crucial strategy in this respect. Again, when the Cooke and Wheatstone telegraph had been laid down from Paddington to Slough, it was licensed out to a showman, Thomas Home, who charged the public a shilling a go for the novelty of sending messages of all kinds down the line. Home advertised special events – like the playing of a game of chess over the telegraph – to advertise his exhibit. As well as being a potentially useful invention, the telegraph was important for practical electricians because it provided a way of exhibiting the mysterious power of electricity. It was an agent that 'far exceeds even the feats of pretended magic and the wildest fictions of the East' and would bring about the 'annihilation of time and space'.

Electrical exhibition and invention held out the prospect of affordable luxury to the middle classes. As Alfred Smee boasted in his *History of Electro-metallurgy*, 'At present a person may enter a room by a door having finger plates of the most costly device, made by the agency of the electric fluid. The walls of the room may be covered with engravings, printed from plates originally etched by galvanism, and multiplied by the same fluid. The chimney piece may be covered with ornaments made in a similar manner: At dinner the plates may have devices given by electrotype engravings, and his salt spoons gilt by the galvanic fluid.' Electroplated goods like the ones Smee describes here were prominently on display at the Great Exhibition as examples of electricity's power. Electricity held out the prospect of being able to replace the hand of man. Even Michael Faraday, not a great exponent of electricity's utility as a virtue in its own right, wondered at 'what man may do, now that Dame Nature has become his drawing mistress', and William Robert Grove looked forward to the day when, 'instead of a plate being inscribed, as "drawn by Landseer, and engraved by Cousins", it would be "drawn by Light, and engraved by Electricity!"'

New inventions were prized according to the ways in which they could be exploited to produce a striking exhibition. In 1849, for example, Edward Staite, anxious to display the virtues of his newly invented electric arc-light, put on a spectacular show in Trafalgar Square: 'The apparatus was so placed ... as to illuminate the whole of Trafalgar-square, the rays reaching as far as Northumberland-house ... The rays were continually moved, and as they swept through the foggy atmosphere, they produced the same sort of illumination as the sunlight through atoms of dust. The objects upon which they fell were most brilliantly lighted. The Nelson column, which was selected as the principal point, being frequently as conspicuous as noonday. If the illumination can be sustained, there is no other means of lighting the streets that can at all be compared with this electric light.' Providing a virtuoso performance like this was integral to the electrical inventor's success, not only in attracting investors who might financially support his invention, but in providing himself and his product with a public image. Performance and showmanship were part of the process of self-fashioning for practical electricians. Staite even commissioned a special ballet, *Electra*, that performed to rave reviews at Her Majesty's Theatre in May 1849. The star of the show was his electric light, which the ballet had been specifically produced to demonstrate.

The great hope for electrical invention, as Henry Noad had indicated, was the replacement of steam power by electricity. Indeed, steam had barely been established as a viable source of locomotive power during the 1830s before hopeful commentators were enthusiastically prophesying its imminent demise. Projects and proposals for electrical locomotion were therefore a staple part of exhibition culture. When the self-publicising American blacksmith Thomas Davenport and his promoters brought a model of his patent electromagnetic engine over the Atlantic, they did not just show it off to the favoured few at the London Electrical Society. It was put on show to great acclaim at the Adelaide Gallery as well. Commentators were confident that electrical power was the future. 'I am free to confess that I cannot discover any good reason why the power may not be obtained and employed in sufficient abundance for any machinery', declaimed the *Morning Herald*'s correspondent, 'why it should not supersede steam, to which it is infinitely preferable on the score of expence [*sic*], and safety, and simplicity ... Half a barrel of blue vitriol, and a hogshead or two of water, would send any ship from New York to Liverpool; and no accident could possibly happen, beyond the breaking of the machinery, which is so simple that any damage could be repaired in half a day.' Electrical inventors like the Glaswegian scientific instrument-maker Robert Davidson put model electrical locomotives on show at exhibition halls in London and elsewhere.

Throughout the 1840s and beyond, electrical exhibitions fuelled a fascination for electricity. Polite natural philosophers like Michael Faraday tended to be distinctly sniffy about this kind of commercial activity. Even Faraday's counterpart at the London Institution, the Welshman William Robert Grove, who was rather more sympathetic towards the utilitarian approach than his colleague at the Royal Institution, rather looked down his nose at what was going on. 'It would scarcely add to the dignity of philosophy, or to the reverence due to its votaries, to see them running with their various inventions to the patent office', he complained. 'If parties look to money as their reward, they have no right to look for fame; to those who sell the product of their brains, the public owes no debt.' Grove, it should be noted, however, was not complaining about utility as such – he was complaining about the profit motive. Like many of his fellow gentlemen of science (and Faraday), he was trying to draw firm demarcations between discovery and invention, between philosophy and industry. Men like Sturgeon and his fellow practical electricians had no truck with these nice distinctions. They had their own ideas about what electricity was all about, and what kinds of activities were appropriate for its practitioners. In their view, they were quite entitled to the kudos of discovery – and the cash of invention if they could manage it, as well.

Part VI: The Great Experimenter

· Chapter 20 ·

An Experimental World

Natural philosophers like Faraday regarded experiment as being at the very core of what they did that made them men of science. British scientific practitioners at the beginning of the 19th century often looked back to the Elizabethan courtier and lawyer Sir Francis Bacon as the inspiration for their natural philosophical approach. In their view, he had established the inductive method – the notion that the best way of finding out about nature was through diligent and disciplined observation and experiment, rather than abstract theorising. In 1830, at the beginning of what would turn out to be Faraday's most productive decade of experimental discovery, the astronomer John Herschel – son of the discoverer of Uranus – published his *Preliminary Discourse on the Study of Natural Philosophy*. Herschel regarded himself as being very much in the Baconian tradition, wedded to the idea that disciplined experimentation was at the very core of sound scientific method. Faraday would have agreed with him entirely. He was distrustful of abstract theorising – like many of his British contemporaries.

One of the reasons why Faraday was uncomfortable with the Frenchman André-Marie Ampère's electrical researches during the 1820s was that they were couched in the abstract mathematical language of contemporary French physics. Faraday thought that mathematics was a distraction from the real world of experiment – a problem that he would also have with James Clerk Maxwell's work a few decades later. It should be clear, however, that the term 'experiment' covers a variety of sins. Sturgeon also argued that the science of electricity was all about experiment. What Faraday – or, for that matter, Herschel – thought experiment was about, was very different from the vision offered by Sturgeon and his fellow members of the London Electrical Society.

Contemporary accounts of Faraday in the laboratory emphasise the care he took to ensure proper order, and the control he exercised over his assistants. Faraday's particular assistant, Sergeant Anderson – who had first come to work for him while Faraday was carrying out experiments on improving optical glass for the Royal Society – was kept on 'simply because of the habits of strict obedience his military service had given him'. As one of Faraday's early biographers noted, he was nothing if not methodical in his experimental arrangements. Before starting on a new bout of research, Faraday 'would give Anderson a written list of the things he would require at least a day before – for Anderson was not to be hurried. When all was ready, he would descend to the laboratory, give a quick glance to see that all was right, take his apron from his drawer, and rub his hands together as he looked at the preparations for his work. There must be no tool on the table but such as he required ... He would put away each tool in its own place as soon as done with, or, at any rate, when the day's work was over, and he would not unnecessarily take a thing from its place.' John Tyndall thought that Faraday's 'sense of order ... ran like a luminous beam through all the transactions of his life'.

From the beginning of the 1830s onwards, Faraday was certainly carefully systematic in his experimental work to a degree that few predecessors had been. He started keeping an experimental Diary in which he diligently recorded the results of each day's labours in the Royal Institution's basement laboratory. When he published the first of the series of *Experimental Researches in Electricity* that, over the next few decades, would cement his stellar reputation as one of Europe's greatest experimenters, he divided that first presentation into a sequence of numbered paragraphs. That sequence of numbered paragraphs continued unbroken through to paragraph 3,242 at the end of the 29th series of *Experimental Researches* in 1851. The notes in his Diary were similarly numbered, ending at

Illustration 11: Michael Faraday's laboratory at the Royal Institution. The stairway up to the lecture theatre is visible on the right.

paragraph 16,041. We cannot know the extent to which this strategy was worked out in Faraday's mind when he started work on what would become the first series, read to a meeting of the Royal Society on 24 November 1831. In fact, in his preface to the first volume of *Experimental Researches*, covering series one to fourteen, Faraday reminded his readers that 'it was not written as a *whole*, but in parts; the earlier portions rarely having any known relation at the time to those which might follow'. Nevertheless, he was clearly making a certain statement about how he regarded his experimental work and the thread he saw running through it, by organising it in this way.

John Tyndall, again, remarked that 'Faraday entertained the opinion that the discoverer of a great law or principle had a right to the "spoils" – that was his term, arising from its illustration'. This at least partly explains the way in which the *Experimental Researches* were presented. Faraday, after all, had had his fingers burnt over the issue of intellectual property before, in the controversy surrounding his electromagnetic rotation experiments. He was not about to allow that to happen again. We shall see over the next few pages the care he took to ensure his own property rights over his experimental discoveries, and the sensitivity with which he responded to what he regarded as breaches of those rights. As Tyndall also noted, Faraday was 'a man of excitable and fiery nature; but through high self-discipline he had converted the fire into a central glow and motive power of life, instead of permitting it to waste itself in useless passion ... Faraday was *not* slow to anger, but he completely ruled his own spirit, and thus, though he took no cities, he captivated all hearts.' The theme of iron self-discipline is a constant in Victorian accounts of Faraday's character. It is also the key to understanding his success as an experimenter. It was how he ensured the success of his experimental work's trajectory from the basement laboratory of the Royal Institution to the gentlemanly world outside.

· CHAPTER 21 ·

INDUCTION

According to his experimental Diary, Faraday started work on the series of experiments that would lead to his discovery of electromagnetic induction on 29 August 1831. He had already ordered the preparation of a soft iron ring wrapped with two separate coils of insulated copper wire. One coil of wire was connected to a galvanic battery. The other was a continuous loop passing over a magnetic needle, which Faraday hoped would detect any current of electricity passing through the coil. The aim of the experiment was simple. Faraday himself in 1821 had played an important role in demonstrating the relationship between electricity and magnetism. Ampère, in France, had shown that coils of current-carrying wire could act like magnets. Faraday, like many others at the time, was looking for the reverse effect. He was trying to produce electricity from magnetism. According to his Diary, as soon as he hooked the first coil up to his battery he noticed a 'sensible effect' on the magnetic needle, which twitched briefly at the moment when electricity first started flowing through the first coil. The same happened, although in the opposite direction, at just the moment the battery connection was broken. He carried on experimenting for the rest of the day, finally noting in his Diary that the curious effect was 'due to a wave of electricity caused at moments of breaking and completing contacts'.

It seems fairly likely that Faraday already knew what he was looking for. Why else, after all, would he have ordered the preparation of his iron ring and coil apparatus? The arrangement bears a superficial similarity, at least, to an electromagnet with the two arms meeting to form a continuous ring. Since the electromagnet's invention a few years earlier, Gerrit Moll in the Netherlands and Joseph Henry in the United States had greatly improved Sturgeon's original instrument. Maybe Faraday had been playing with electromagnets as well and had noticed something peculiar happening to the coil on one arm when the coil on the other arm was attached to a battery. After that first flurry of activity, Faraday made no more notes in his Diary for almost a month, beyond noting on 12 September that he had ordered the preparation of a number of coils. He started work in earnest again on the 24th. He was trying hard in particular to find different ways of demonstrating the existence of electricity in the secondary coil. He tried unsuccessfully to find some evidence of the chemical and heating effects of electricity in the coil, but did succeed in getting a 'very distinct though small' spark to appear on 1 October. This was tedious, difficult and repetitive work. The kinds of effects that Faraday was looking for were tiny and at the limits of his equipment's capacity to detect reliably.

Faraday's Diary and his presentation to the Royal Society show how he used the Royal Institution's resources and his privileged position there to good effect in pursuing his researches. Time-consuming and boring tasks like winding wire coils could be relegated to his assistants. Samuel Hunter Christie, Professor of Natural Philosophy at the Woolwich Royal Military Academy and Secretary of the Royal Society, lent him the Society's massive magnet to help magnify the delicate effects he was trying to isolate. In his presentation to the Royal Society, eventually published in the Society's *Philosophical Transactions*, Faraday described two kinds of induction: volta-electric and magneto-electric. Volta-electric induction was produced when a coil of wire was wrapped around an inner coil that could be connected to a battery. A current was produced in the outer coil whenever contact between the inner coil and the battery was made or broken. Magneto-electric induction could be produced in two ways. The first was by using the ring apparatus with which he had first started experimenting. The second method was simply by moving a bar magnet

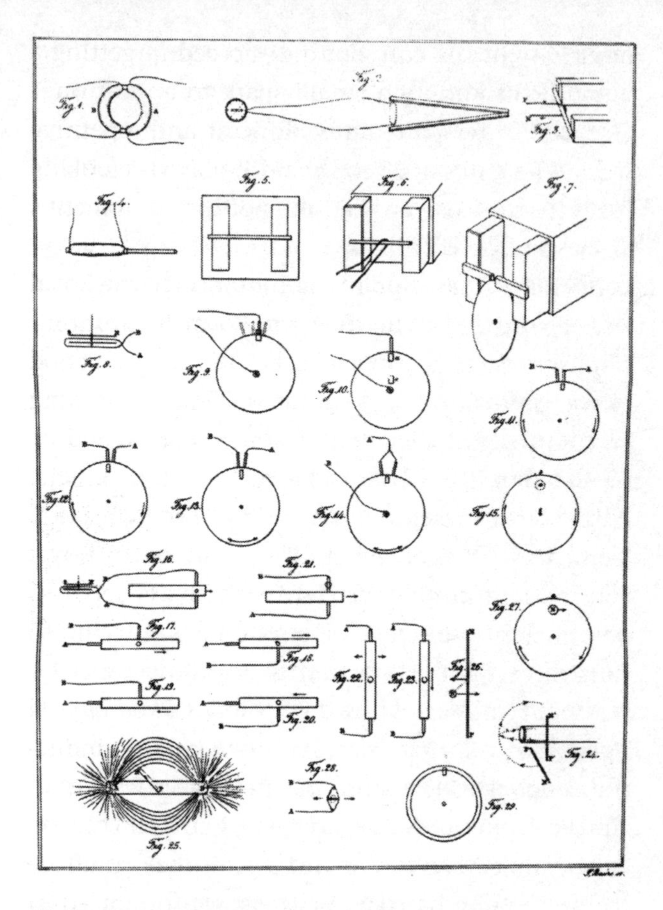

Illustration 12: Diagrams of some of Faraday's experiments that led to the discovery of electromagnetic induction.

in and out of the coil of wire. A current was produced in the coil depending on the direction in which the magnet was moved. The Royal Society's large magnet was used to show how electricity could be produced from magnetism by spinning a copper disc between its poles.

Faraday's discovery of the evolution of electricity from magnetism caused quite a splash in scientific circles, and he clearly felt that it constituted a major coup. A few days after his communication had been read before the Royal Society he was writing triumphantly to his friend Richard Phillips, describing how 'egotistical' he felt at having discovered phenomena that had eluded so many 'great names' in the past. He also dashed off a letter to Jean Nicolas Hachette, an old Parisian acquaintance from the days of his Grand Tour with Davy. Hachette promptly passed the letter on to François Arago, the permanent secretary of the Royal Academy of Sciences who, just as promptly, had it read out at the Academy's next meeting. The upshot of all this was that two Italian natural philosophers, Leopoldo Nobili and Vincenzo Antinori, succeeded in repeating the induction experiments and published their own results – before Faraday's own account was published in the *Philosophical Transactions*. Faraday was furious. His own unfortunate experiences over the past decade had left him extremely sensitive on matters of priority and propriety in scientific discovery. In this case Faraday felt that both Hachette and the two Italian philosophers had overstepped the mark. In their enthusiasm they had stolen his credit. Hachette had acted dishonourably by making a private communication public, and the Italians were no better than plagiarists.

Faraday wrote angrily to Hachette expressing his displeasure. Hachette's response was a masterpiece of Gallic bafflement in the face of what he clearly regarded as a peculiarly English *amour propre*. He assured Faraday that the presentation he had made to the Royal Society, along with the communication that had been read out at the Royal Academy of Sciences in Paris, left no one in any doubt at all about who had the priority in this discovery. After all, the two Italians themselves made it quite explicit in their own publication that they were merely repeating and extending Faraday's work. Faraday refused to be mollified. He persuaded the editors of the *Philosophical Magazine* to publish a translation of Nobili and Antinori's paper and added a note of his own, making his complaint public. 'I may perhaps be allowed to say ... that had I thought that that letter to M. Hachette would be considered as giving the subject to the philosophical world for general pursuit, I would not have written it; or at least not until after the publication of my *first paper*.' Hachette clearly felt that this was getting ridiculous. 'You announced a great discovery to the Royal Society; it was transmitted to France in your name; all the glory of the inventor was assured you; what more could you desire than that?', he shrugged; 'that no one else should work in the mine that you have opened; but that's impossible.' The problem, of course, was that that was exactly what Faraday wanted.

Faraday wanted his 'grande découverte' (as Hachette flatteringly termed it) to be his own property. He wanted the intellectual spoils from it for himself. Just how difficult that task would be became clear over subsequent months and years. It was not long before not only Nobili and Antinori in Italy, but also David Forbes, Professor of Natural Philosophy at Edinburgh, succeeded in repeating and extending Faraday's experiments. Where Faraday had succeeded in producing an electric spark by induction using an electromagnet, Nobili and Antinori and Forbes laid claim to having been able to produce the same effect using an ordinary magnet. Faraday's discovery was soon embodied in an instrument as well – and an instrument made by a Frenchman rather than an English instrument-maker, to boot. Hippolyte Pixii came up with a contrivance in which coils of wire rotated between the poles of a magnet could be made to produce a continuous current that could be used, among other things, to decompose water. Joseph Saxton, the Adelaide Gallery's in-house instrument-maker, also came up with a similar machine and demonstrated it at the Gallery before showing it off at the British Association for the Advancement of Science's annual meeting in Cambridge in 1833. There, 'the ingenious invention of Mr. Saxton, by which the transient electrical currents might exhibit their effects in so brilliant and so powerful a manner' was hailed by no less a personage than the poet Samuel Taylor Coleridge as being on a par with Faraday's own discovery.

That bombastic piece of hyperbole produced a swift reprimand from the *Literary Gazette*, accusing Saxton of stealing other people's discoveries. Saxton just as swiftly penned an indignant response, denying that he had ever laid claim to anyone else's intellectual property. In particular he denied having copied his version of the magneto-electric machine from Pixii (whose machine he claimed never to have heard of). With the help of instrument-maker Francis Watkins, who owned one of the Pixii machines, Saxton soon organised a grand competition at the Adelaide Gallery between his own machine and Pixii's. Eminent witnesses – including Faraday himself – were invited along to see the two machines put through their paces. As the *Literary Gazette* itself reported, the Pixii machine, operated by Watkins, 'very rapidly decomposed water, first in a single tube, hydrogen gas being evolved from one pole of the armature while oxygen gas was given off at the other ... We were then favoured with an experiment quite new in this country, namely, that of charging the Leyden jar with magneto-electricity; the truth of which was made evident by the aid of a delicate electroscope, the gold leaves of which very sensibly diverged.' When Saxton's machine took to the stage, 'it gave powerful shocks, brilliant sparks, heated a platinum wire red hot, and decomposed water'. The *Gazette* had no doubt in awarding the laurels to Saxton, whose 'splendid apparatus attracted the universal admiration of the scientific company present, not only from the beautiful and extraordinary effects produced by it, but also from its very superior mechanical arrangement'.

The story shows just how difficult it was for Faraday to prevent his discovery running away from him. As far as Faraday himself was concerned, his induction experiments mattered because they were evidence of what he called the 'electrotonic state' – a peculiar condition of matter which appeared to manifest itself only during induction, and which Faraday suspected of being 'a very important influence in many if not most of the phenomena produced by currents of electricity'. Faraday thought that the mere 'discoverer of a fact' was as nothing compared with 'he who refers all to still more general principles'. That was what the idea of the electrotonic state promised, and he felt that as its discoverer he had – or ought to have – the exclusive right to its further elucidation. Others, however, had quite different ideas about why induction mattered. Neither Pixii nor Saxton, to be sure, had the slightest interest in Faraday's electrotonic state. What induction promised to them was the opportunity to invent new instruments that could produce more spectacular and thrilling displays of electrical effects. To other hopeful inventors like William Sturgeon, Joseph Henry, Thomas Davenport and E.M. Clarke, to name but a few, it also offered the prospect of making electromagnetism into a viable means of locomotion that could challenge steam. Faraday's 'grande découverte' was, in practice, very quickly picked up, re-appropriated and assimilated into a whole range of different experimental cultures.

· CHAPTER 22 ·

FURTHER EXPERIMENTAL RESEARCHES

Faraday did not allow the contretemps surrounding his discovery of induction and the electrotonic state to distract him from continuing his experimental researches, however. On 12 January 1832 he delivered the Royal Society's prestigious Bakerian Lecture and took advantage of the opportunity to present material that would become his second series of *Experimental Researches*. In the Bakerian Lecture he presented experiments aimed at discovering whether electrical currents could be induced using the Earth itself instead of an ordinary magnet. His experiments had been successful, showing that electric currents could, indeed, be induced by the Earth's magnetism. When a coil of wire wound around a soft iron core was held so that it lay along the direction of magnetic dip and then quickly turned around so that it pointed in the opposite direction while still lying along the angle of dip, a galvanometer needle was immediately deflected, showing the passage of a transient current. Faraday proceeded to suggest ways in which this phenomenon could be used to explain various features of terrestrial electricity such as the existence of subterranean currents of electricity – an issue close to William Sturgeon's heart, as we have seen. In his third series of *Experimental Researches*, Faraday temporarily abandoned his work on induction to establish the identity of electricity from different sources and to investigate the phenomenon of electro-chemical decomposition. In the process he succeeded in provoking Sturgeon into a direct attack on his reputation and experimental competence.

Surprisingly, maybe, the question of the identity of electricity from different sources (such as common electrical machines, galvanic batteries, thermo-electricity, and even electric eels) was still a matter of occasional debate in the 1830s. One reason for this was that the effects seemed so varied. Common electrical machines were good for producing huge sparks and shocks, for example, but useless at producing magnetic effects such as deflecting a needle. In his third series, read to the Royal Society on 10 and 17 January 1833, Faraday tried to overcome this difficulty by devising arrangements of the different kinds of methods for producing electricity such that they all produced roughly equivalent effects. Having established the identity of the electricities to his satisfaction, Faraday proceeded to investigate what he called the 'relation by measure of common and voltaic electricity'. This was an effort to find a common way of measuring electricity, regardless of its source, as a way of further demonstrating the identity of the electricities and as a way of establishing general principles. In the fifth series he established the law of electro-chemical decomposition which says that, as he put it, 'for a constant quantity of electricity, whatever the decomposing conductor may be ... the amount of electro-chemical action is also a constant quantity'. In the seventh series, he suggested (on the basis of his law of chemical decomposition) that the decomposition of water during the passage of electricity could be used as an absolute measure of the quantity of electricity passed.

Sturgeon mounted a fierce attack on Faraday's conclusions in the pages of his *Annals of Electricity*. What he has to say is particularly interesting since it makes explicit the differences between his and Faraday's approaches to electrical experimentation. By and large, Sturgeon simply denied that Faraday's experiments showed what he said they showed. He took particular exception to Faraday's suggestion that a single instrument – what Faraday called a voltameter – could be used as an absolute measurer of electricity. Sturgeon argued that different sources of electricity were better or worse at producing any given effect. The fact that, for example, a particular battery decomposed more water in a voltameter than another battery, did not mean that it would produce a better spark, or provide more power for an electromagnet. He suggested that what Faraday called a voltameter would be better described as an electro-gasometer since all it could really be said to measure was the amount of gas given off during the decomposition of water by electricity. This was a difference in perspective about what such experiments were for. Faraday was interested in absolute measurement. Sturgeon, on the other hand, simply wanted to find ways of assessing the capacity of different kinds of sources of electricity to produce particular kinds of effect. Sturgeon's implication was that it was only Faraday's reputation that allowed him to make the kind of experimental claims he did. 'The errors of a fashionable man', sneered Sturgeon, 'whatever may be the nature of his pursuits, are almost sure to lead those astray who have either no desire or no opportunity to judge for themselves, and there is not, perhaps, amongst the numerous errors into which Dr. Faraday has fallen, one more eminently calculated to mislead the unwary experimenter than the pretended accuracy of an instrument, the principles of which, he either neglected to reveal, or of which he had not the slightest knowledge.' Faraday did not rise to Sturgeon's bait – at least not in public. In private, he acknowledged that he 'had seen Mr Sturgeon's criticism but it did not disturb me simply because I was glad to find that the arguments he uses are of no force ... It would be some fun to send him Baron Humboldts letter to me in which he selects that very paper as the foundation of compliment and praise only I cannot consent to use the letter for such a small purpose or quote Humboldts name in sport.'

Faraday's ninth series of *Experimental Researches*, first read to the Royal Society on 29 January 1835, returned to the topic of induction. At the start of his published paper he acknowledged that this work on 'the induction of a current on itself' was instigated by an observation reported to him by a Mr Jenkin. He later remarked that the 'number of suggestions, hints for discovery, and propositions of various kinds, offered to me very freely, and with perfect goodwill and simplicity on the part of the proposers for my exclusive investigation and final honour, is remarkably great, and it is no less remarkable that but for one exception – that of Mr. Jenkin – they have all been worthless'. Jenkin had noticed that an electric shock could be produced at the moment contact was broken from a battery consisting of a single pair of plates (something usually requiring a far more powerful apparatus) if there was a helix (a coil of wire) in the circuit. Faraday interpreted this as a new kind of induction and further evidence for the existence of his electrotonic state. It was an example of a current inducing a secondary current in its own circuit. As he put it, the 'strong spark in the single long wire or helix, at the moment of disjunction, is therefore the equivalent of the current which would have been produced in a neighbouring wire if such second current were permitted'. He demonstrated this by showing that if a neighbouring wire was present, then a secondary current was induced in it and no extra current appeared in the original – or primary – circuit.

The ninth series provoked another blast from Sturgeon in the pages of his *Annals of Electricity*. Sturgeon, too, had been inspired to examine the phenomenon of a spark produced from a single pair of plates, after having been told about it by a Mr Peaboddy, a visitor at the Adelaide Gallery who had heard about the American experimenter Joseph Henry's investigations of it (which actually preceded Faraday's own experiments). Where Faraday had interpreted the experiment in the light of his own work on induction, Sturgeon saw it as an example of his own efforts to construct more economical ways of displaying electrical phenomena. It was a way 'to convert *quantity* of the electric fluid into *intensity*', and hence produce more powerful phenomena. He interpreted the effect as being due to the momentum of the electric fluid flowing back and forth through the wire. He condemned Faraday's explanations as being based on insufficient experiments and the use of inappropriate instruments such as the galvanometer. His results, he said, were 'the very reverse of those which Dr. Faraday has drawn from the *partial*, and consequently *inconclusive* character of his experimental data; but they are precisely such as might have been expected by any one conversant with the nature of electro-dynamic action'. We do not know what Faraday made of this second attack. He certainly made no effort to respond to it in public.

In 1845, fourteen years after he had started his *Experimental Researches*, Faraday communicated the nineteenth series to the Royal Society on 20 November, on 'the magnetization of light and the illumination of magnetic lines of force'. These experiments were the result of his conviction 'that the various forms under which the forces of matter are made manifest have one common origin; or, in other words, are so directly related and mutually dependent, that they are convertible, as it were, one into another, and possess equivalence of powers in their action.' They were Faraday's effort to find a link between magnetism and light to add to the links he had found between magnetism and electricity. As Faraday reported to the Royal Society, he found that when polarised light was passed through heavy optical glass, 'if a magnetic line of force be *going from* a north pole, or *coming* from a south pole, along the path of a polarized ray coming from the observer, it will rotate that ray to the right-hand; or, that if such a line of force be coming from a north pole, or going from a south pole, it will rotate such a ray to the left-hand.' He found similar effects when the polarised rays were passed along the axis of a wire helix carrying an electrical current, depending on the direction in which the current passed.

As Faraday wrote to William Whewell, he was anxious to keep a lid on the new discovery until he had explored its ramifications. He clearly regarded the nineteenth series as being on a par with the first series in terms of significance. He was mindful of the brouhaha with Hachette and the Italian philosophers the last time round, and wanted no repetition. He begged Whewell to be discreet: 'But do not say, even, that you are aware I am so engaged. I do not want men's minds to be turned to my present working until I am a little more advanced.' He was a little late in his warning. Whewell had contacted Faraday asking for more details after reading an account of his magneto-optic experiments in the *Athenaeum* magazine, almost two weeks before Faraday's paper was to be read to the Royal Society. John Herschel contacted Faraday as well after reading the report, reminding him that he had tried such experiments unsuccessfully himself as far back as 1825 – not something that Faraday, mindful of priority, wanted to be reminded of. Faraday's response to Herschel's pestering for more details of the experiments was to send him a sealed note, to be opened only with his permission, outlining the discovery. He was not going to let this one get away from him. Neither Herschel nor Whewell were that impressed when they eventually got the details they wanted. Herschel thought Faraday had found a magnetic effect on the crystalline structure of the optical glass he had used, rather than a direct link between magnetism and light. Whewell scoffed to David Forbes in Edinburgh about the 'overcharged importance of Faraday's view of his recent discoveries'.

Faraday produced another ten series of *Experimental Researches* following his discovery of the magneto-optic effect. Many of these series were devoted to further working out the implications of his magneto-optic discoveries. He expended several series to investigating the magnetic properties of a whole range of different substances, from metals and crystals to gases and even empty space. In the 24th series, delivered in 1850, he outlined some experiments designed to investigate the relationship between electricity and gravity. Again, it was the 'long and constant persuasion that all the forces of nature are mutually dependent, having one common origin, or rather being different manifestations of one fundamental power' that led him to 'think upon the possibility of establishing, by experiment, a connexion between gravity and electricity, and so introducing the former into the group'. That the experiments were unsuccessful did nothing to dispel his 'strong feeling of the existence of a relation between gravity and electricity, though they give no proof such a relation exists'. The *Experimental Researches* – all 29 series of them – were, by any standards, a remarkable achievement. Continued over three decades, they constituted a treasure trove of experimental detail and a record of continuous research that was unparalleled among Faraday's contemporaries. They also showed just how different Faraday's ideas of experiment and experimental discipline were from those of other electrical experimenters in Britain at the time.

· Chapter 23 ·

Lines of Force in Space

To fully make sense of the ways in which Faraday as an experimenter differed from practical electricians like Sturgeon, we need to appreciate the broader context of his experimental life. Faraday's aim when performing in front of his genteel audience at the Royal Institution was to make them feel that they were in 'Nature's school'. He wanted them to leave his lectures feeling as if they had been in direct communion with nature – that she had been their teacher – rather than as if the lessons had been conveyed by someone else. Faraday wanted to reduce nature to a series of facts as well. As he noted himself on more than one occasion, he was obsessed by facts. 'I was never able to make a fact my own without seeing it', he wrote to a friend; 'if Grove, or Wheatstone, or Gassiot, or any other told me a new fact & wanted my opinion, either of its value, or the cause, or the evidence it could give in any subject, I never could say anything until I had seen the fact.' This was what he wanted his audience at the Royal Institution to see as well – the facts of nature unadorned by any embellishment.

The same urge is clear in the *Experimental Researches* also. They were aimed at the making and presentation of a connected series of facts.

This is reminiscent of William Whewell's comment that we encountered earlier. Whewell remarked in his magisterial *Philosophy of the Inductive Sciences* that, in Faraday's researches, 'all the superfluous and precarious parts gradually drop off from the hypothesis ... and the abstract notion of polarity – of equal and opposite powers called into existence by a common condition – remains unincumbered by extraneous machinery'. Faraday worked hard at making his own role in producing the facts he presented to his audiences (either at the Royal Institution's lecture theatre or in the pages of the *Philosophical Transactions*) as invisible as possible. He also worked hard at making the experimental apparatus with which he produced those facts as unobtrusive as possible. Even as far back as his early youthful discussions with Benjamin Abbot on the art of lecturing, Faraday recognised that if what he wanted to show his audience was nature speaking for herself, then he would have to draw attention away both from himself and from the machinery he used to make the effects. In this sense, the way Faraday performed and presented his experimental work fitted in rather well with the kind of idealism espoused by William Whewell – that spokesman for genteel and moderate

Anglicanism. It was a way of performing that suited his equally genteel and Anglican audience at the Royal Institution as well. They did not want to be distracted by messy machinery any more than Faraday wanted them to be.

Just as Faraday tried to make it appear to his audiences that neither he nor his experiments were really there at all, so the view of nature that he offered was one in which the instruments did not really matter either – paradoxically for such a superlative experimenter. Where nature for William Sturgeon, for example, was made up of machines just like the ones he experimented with and demonstrated to his audiences, Faraday's nature was made up of disembodied lines of force in space. During the 1830s and 40s, Faraday developed a view of electricity and the relationship between electrical forces and matter which suggested that everything interesting went on in the spaces surrounding his experimental apparatus rather than in the apparatus itself. Most of Faraday's contemporary electrical experimenters generally subscribed to the view that electricity was some kind of fluid. It flowed through wires in much the same way that water flowed through a pipe. Faraday disagreed. Electrical and magnetic forces were to be found in the spaces surrounding the apparatus – what he eventually started to call electric or magnetic fields. His views on the electrotonic state in his work on induction were an early expression of this notion. He represented electrical and magnetic fields as lines of force emanating from charged particles of electricity or from the poles of a magnet. It was these physical lines of force – regarded by Faraday as just as real as solid matter – that were really significant for understanding the nature of electricity or magnetism. In some ways, for Faraday, what was usually thought of as matter and its effects was only a convenient shorthand for talking about the properties of these lines of force.

Very few indeed of Faraday's contemporaries took particularly seriously these rather peculiar views about lines of force in space. They were probably what Whewell had in mind when he disparaged Faraday's high opinion of his theoretical views to his friend David Forbes, as mentioned earlier. His views about the mutual convertibility, or mutual origins, of the different forces like electricity, heat, light or magnetism were rather more respectable and widely shared. In 1846, Faraday's friend, fellow experimenter and opposite number at the London Institution, William Robert Grove, published the *Correlation of Physical Forces*, arguing that all the forces of nature were mutually correlated and dependent on each other, 'that the various imponderable agencies, or the affections of matter which constitute the main objects of experimental physics, viz. Heat, Light, Electricity, Magnetism,

Chemical Affinity, and Motion are all Correlative, or have a reciprocal dependence'. Grove presented many of Faraday's own experimental discoveries – like the mutual relationship between electricity and magnetism and the magneto-optic effect – as examples of correlation. From the vantage point of a few decades later in the 19th century, when the doctrine of the conservation of energy was firmly established, Faraday's *Experimental Researches* could very easily be made to look like an extended exercise in demonstrating the experimental evidence for conservation. As we shall see, it was the way that Faraday was understood by the following generation of electrical scientists that made his views on lines of force seem so significant too.

PART VII:

THE ELECTRICAL CENTURY

· Chapter 24 ·

Worlds Apart

During the early 1840s, Faraday suffered a series of what we might now think of as mental or nervous breakdowns. He complained of anxiety, headaches, lapses in concentration and memory loss. One of Faraday's biographers, his friend and colleague at the Royal Institution Henry Bence Jones, noted of the year 1841 that 'Loss of memory and giddiness had long, occasionally, troubled Faraday, and obliged him to stop his work. But now they entirely put an end to his experiments.' Bence Jones even divided Faraday's experimental researches into two periods – one before and one after his breakdown. John Tyndall, another close friend and biographer, attributed Faraday's condition to the cumulative effect of 'the mental strain to which he had been subjected for so many years'. Part of Faraday's problem – as his biographers recognised – was his tremendous workload. On top of his prodigious experimental output, Faraday was responsible for the day-by-day administrative business of the Royal Institution, as well as organising and delivering lectures and Friday Evening Discourses. Neither of them mentioned another possible source of conflict, however – that Faraday was increasingly feeling the strain of reconciling his private Sandemanian religion with his public persona. In 1840, Faraday had been elected an Elder of the Sandemanian church – a position that caused him a great deal of anxiety as well as satisfaction. Disastrously, in 1844 he was briefly excluded from the church following conflict within its membership. Even following his reconciliation, Faraday remained worried that a second exclusion would lead to his being permanently barred from his religious community. For a time in 1850 it seemed as if that second and final exclusion might well come about, and the threat of ultimate excommunication was enough to leave Faraday teetering on the brink of nervous breakdown once more.

By the 1840s, Faraday was at the peak of his public and scientific reputation. He was, as we have seen, widely regarded as England's greatest natural philosopher. His lectures at the Royal Institution drew huge crowds of fashionable acolytes. He was certainly the country's foremost exponent of science to a popular audience. With this popularity, along with the mastery of nature demonstrated in his painstaking experimental researches, came a litany of honours. By the 1840s, as well as being a Fellow of the Royal Society, Faraday was a member of most of the learned societies in Europe. He had been a Corresponding Member of the Academy of Sciences in Paris since 1823. Over the years, numerous other honours were added to the list, including membership of the Imperial Academy of Sciences in St Petersburg, the Royal Society of Sciences in Copenhagen, the Royal Academy of Sciences in Berlin and the American Academy of Arts and Sciences. In 1832 he had been awarded the degree of DCL (Doctor of Civil Law) by the University of Oxford – an almost unheard-of distinction for a working-class lad and a dissenter to boot. Faraday famously never accepted a knighthood, unlike his master Humphry Davy. He also turned down another honour that Davy had accepted. In 1858 he was offered – and refused – the Presidency of the Royal Society. In 1864, only a few years before his death, he was also offered – and refused – the Presidency of the Royal Institution.

One outcome of Faraday's growing reputation was that he was increasingly called upon by government as a scientific expert. During the late 1820s he had been commissioned by the Royal Society to carry out some experimental work on improving optical glass – an area where it was felt that English glass-makers were falling behind their Continental counterparts. In 1836 he was appointed Scientific Adviser to Trinity House, the body responsible for the care and upkeep of lighthouses in British waters. In this capacity, Faraday was responsible for examining proposals concerning things such as the introduction of new lighting systems or new methods of ventilating lighthouses. Among other things, he was responsible during the early 1860s for examining the possibility of introducing electric lighting systems into lighthouses. In 1844, in the aftermath of the disastrous Haswell Colliery explosion in which 95 people died, Faraday, along with the geologist Sir Charles Lyell, was asked to attend the inquest established to examine the causes of the disaster and report to the Home Office. During the 1850s Faraday was also consulted about ways of trying to improve the quality of Thames water, which was increasingly regarded as a major health hazard in the metropolis. A famous cartoon in *Punch* shows a fastidious Faraday leaving his calling card for a filthy and bedraggled Father Thames. Activities like these were symptomatic of the emergence of a new role for men of science during the Victorian period – as public experts who could be called upon to place their expertise at the service of the state.

These public activities sat uneasily with Faraday's Sandemanianism. His religious practice was a very private affair. It certainly informed his scientific practice through and through. Faraday regarded his task as a natural philosopher as being to read the Book of Nature in just the same way that a good Christian should read the Book of God – that is to say, literally. His belief in such doctrines as the conservation of force was predicated on his belief that only God could destroy those things (such as force and matter) that God had created. However, Faraday's membership of a small and exclusive sect put him at odds with the vast majority of his Royal Institution audience and his fellow natural philosophers. It distanced him from them socially as well as religiously. Sandemanian doctrine taught its adherents to put religious obligation before public duty – one account of Faraday's brief exclusion from the sect in 1844 suggests that it was the result of his obeying a summons to visit Queen Victoria on a Sunday. Other evidence – particularly the fact that a number of other members were excluded from the church at about the same time – suggests that something more was going on. There seems to have been a dispute about church discipline raging across Sandemanianism in Britain and America, and the exclusions in the London meeting-house may well have been part of this broader schism. Faraday, nevertheless, certainly used his Sandemanianism as a way of keeping his distance from the fashionable scientific London that he usually inhabited. It allowed him to stay at arm's length from his genteel audience and fellow philosophers and – paradoxically – preserve his image as the consummate spokesman for nature.

Faraday's self-imposed separation from polite

and scientific society continued into death. He died on 25 August 1867, aged 75, in the 'grace and favour' house in Hampton Court that he had been awarded by Queen Victoria. He left strict instructions that his funeral was to be a private affair for family and close friends. There was to be none of the pomp and circumstance that was increasingly coming to characterise the funeral services of England's great men. He was buried under a simple headstone in the Sandemanian plot at Highgate cemetery. As his niece, Jane Barnard, wrote to Henry Bence Jones, 'The funeral took place on Friday (30th), leaving here at 9.30, and taking up some of the mourners at the Royal Institution, and from thence to Highgate. By my dear uncle's verbal and written wishes, it was strictly private and plain. We could not but follow out his last wishes. I must not lead you to think we did not fully enter into his views, but some would have liked it otherwise.' While being at the very centre of Victorian science and society, Faraday also – and very self-consciously – stood to one side of it too. We will now look at Faraday's 19th-century legacy and examine just what role he really had in the making of the electrical century.

· CHAPTER 25 ·

THE ELECTROMAGNETIC UNIVERSE

In 1855, a young Cambridge graduate, James Clerk Maxwell, read a paper to the Cambridge Philosophical Society 'On Faraday's Lines of Force'. Maxwell had only recently graduated Second Wrangler and first Smith's prizeman at Cambridge and been awarded a prestigious Trinity College Fellowship. He was a mathematician looking for an arena in which to flex his philosophical muscles. On the advice of William Thomson, Professor of Natural Philosophy at Glasgow, Maxwell turned to electricity. Maxwell had asked Thomson how someone who had 'a popular knowledge of electrical show experiments ... ought ... to proceed in reading & working so as to get a little insight into the subject'. Thomson's reply was that Maxwell should try Faraday's researches. In 'On Faraday's Lines of Force', Maxwell picked up on Faraday's suggestion that electric and magnetic forces should quite literally be thought of as lines in space, and incorporated it into a mathematical theory. As he wrote to Faraday, 'You seem to see the lines of force curving round obstacles and driving plump at conductors and swerving towards certain directions in crystals, and carrying with them everywhere the same amount of attractive power spread wider or denser as the lines widen or contract'. Maxwell returned to the theme a few years later with a paper 'On Physical Lines of Force' in which he tried to develop a tangible physical model for the mathematical arguments he had put forward in his original paper.

The eventual result of Maxwell's electromagnetic theorising was the monumental *Treatise on Electricity and Magnetism* (1873), in which Maxwell, by now the newly appointed Cavendish Professor of Physics at the University of Cambridge, tried to establish a fully fledged and comprehensive mathematical theory of electromagnetism. Maxwell's *Treatise* was an immensely dense and complex *tour de force* of mathematical reasoning. It took as its basic assumption that electromagnetic energy was located in a single, space-filling medium – the ether – and set out to provide a mathematical account of the way it interacted with that medium. In it, Maxwell brought together and elaborated his earlier work, putting the flesh on the bones of the electromagnetic ether. As Maxwell had made clear in earlier work: 'In speaking of the Energy of the field ... I mean to be understood literally. All energy is the same as mechanical energy, whether it exists in the form of motion or in that of elasticity, or in any other form. The energy in electromagnetic phenomena is mechanical energy. The only question is, Where does it reside?' The *Treatise* provided a comprehensive account of just where that energy resided, spelling out the mathematical properties of the space-filling ether. It showed, among other things, how waves of electromagnetic energy travelled through the ether at the speed of light, suggesting that light was itself a form of electromagnetic energy.

Faraday's idea of the field was certainly built into the core of Maxwell's electromagnetic theorising. Maxwell himself also certainly made his debt to Faraday's central insight quite clear in all his writings. It takes only a little probing, however, to see that the debt was, in practice, quite limited. Faraday and Maxwell came from very different intellectual traditions and their ideas about how to do physics were accordingly very different too. Faraday had learned his natural philosophy at the feet of Humphry Davy at the Royal Institution. Maxwell, on the other hand, was the product of a rigorous Cambridge training regime that had produced a whole new generation of mathematically expert physical theorists. Faraday never trusted mathematics as a way of doing natural philosophy. In fact, he rather plaintively asked Maxwell whether he could translate his complex equations into a language comprehensible to an ordinary

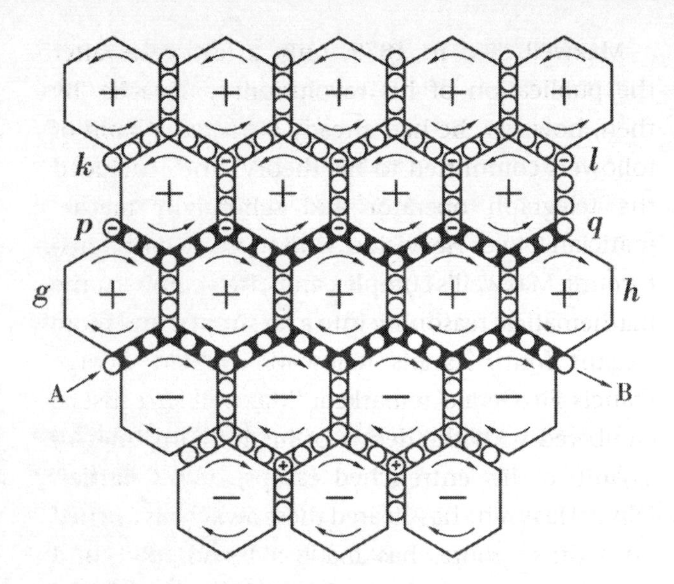

Illustration 13: James Clerk Maxwell's representation of the electromagnetic ether, trying to show how electromagnetic effects might be the results of mechanical movements in the ether.

man. But for Maxwell, mathematics was the only reliable form of natural philosophical reasoning about the physical world. He was at the forefront of a new breed of largely Cambridge-trained mathematical physicists who would completely transform British physics and the science of electricity by the end of the 19th century. Faraday was clearly an important inspiration for this new generation – and a key source of philosophical insight – but he certainly was not one of them.

Maxwell died in 1879, only a few years after the publication of his revolutionary *Treatise*. By then, however, he had already attracted a band of followers committed to his theory. They included the telegraph operator and self-taught mathematician Oliver Heaviside, who succeeded in transforming Maxwell's complex and often cumbersome mathematical reasoning into a far simpler and more elegant form. As his fellow-Maxwellian George Francis FitzGerald remarked: 'Maxwell's treatise is cumbered with the debris of his brilliant lines of assault, of his entrenched camps, of his battles. Oliver Heaviside has cleared these away, has opened up a direct route, has made a broad road, and has explored a considerable trace of country.' FitzGerald himself made major contributions to Maxwellian physics. As far as the Maxwellians were concerned, the key to establishing the truth of Maxwell's theory was to find a way of showing that electromagnetic waves really did travel through the ether as their master suggested. During the early 1880s, FitzGerald had succeeded in calculating the amount of energy such waves would give off. The trick, of course, was to find a way of detecting them. That was the task that the Liverpool professor, Oliver Lodge, set for himself throughout the 1880s. He was pipped at the post in the end, however, by news from Germany.

Maxwell's electromagnetic theory seemed finally

vindicated in 1888 when a young German physicist, Heinrich Hertz, announced that – while working on some experiments with electric sparks – he had found a way of detecting what seemed to be waves of electromagnetic energy: just as predicted by Maxwell's theory. As he explained, 'since the year 1861 science has been in possession of a theory which Maxwell constructed upon Faraday's views, and which we therefore call the Faraday–Maxwell theory. This theory affirms the possibility of the class of phenomena here discovered just as positively as the remaining electrical theories are compelled to deny it. ... as long as Maxwell's theory depended solely upon the probability of its results, and not on the certainty of its hypotheses, it could not completely displace the theories which were opposed to it. The fundamental hypotheses of Maxwell's theory contradicted the usual views, and did not rest upon the evidence of decisive experiments. In this connection we can best characterise the object and the result of our experiment by saying: The object of these experiments was to test the fundamental hypothesis of the Faraday–Maxwell theory, and the result of the experiment is to confirm the fundamental hypothesis of the theory.' His electric waves were the jewel in the crown of Maxwellian physics.

For Maxwellians like Lodge, Hertz's electromagnetic waves were also the final proof – if more proof were needed – of the existence of a space-filling electromagnetic ether. As Lodge put it: 'Persons who are occupied with other branches of science, philosophy, or with literature, and who have not kept quite abreast of physical science, may possibly be surprised to see the intimate way in which the ether is now spoken of by physicists, and the assuredness with which it is experimented on. They may be inclined to imagine it is still a hypothetical medium whose existence is a matter of opinion. Such is not the case.' Maxwell's theory and Hertz's experimental discovery took the ether out of the realm of ideas and made it a fact. The task that many British electrical experimenters and theorists set themselves in the closing decades of the 19th century was not to prove the ether's existence – that had already been done to their satisfaction – but to investigate its structure and properties. They produced ever more elaborate models to demonstrate how electromagnetic effects could be explained as products of the mechanical structure of the ether. This was very far away indeed from Faraday, who thought that his ideas about lines of force in space did away with the need for a space-filling medium like the ether.

· CHAPTER 26 ·

SELLING ELECTRICITY

The Glaswegian physicist William Thomson famously remarked that in order to turn electrical theory into reality, electricity would have to become 'a real, purchaseable tangible object' so that 'we may perhaps buy a microfarad or a megafarad of electricity'. He was suggesting that electricity would be real only when it would be possible to buy a quantity of electricity in much the same way as one might buy a pint of milk. This was making a link between the worlds of natural philosophy and commerce of which Michael Faraday – with his views about the purity of science and the disinterested motives of its practitioners – would have thoroughly disapproved. By the middle of the 19th century, nevertheless – and as we have already seen – electricity was increasingly a commercial proposition. It was on the way to becoming a commodity in just the way that Thomson predicted it would need to do if it was ever going to become part of the real, everyday world. Faraday certainly played a role in turning electricity into a commodity too, but it was by no means a direct one. While he had no objection to the potential utility of his researches, he also regarded utility as being largely beside the point. According to a famous (though possibly apocryphal) story, when Faraday was challenged to explain what use electricity was, his retort was: 'What use is a baby?' Electricity's value, in other words, was not to be defined by its utility. Others, like William Thomson, thought differently.

The first large-scale commercial use of electricity was telegraphy. Several efforts had been made during the first half of the 19th century to find ways of using electricity to carry signals over large distances. It was not until 1837, however, that two Englishmen, Charles Wheatstone and William Fothergill Cooke, took out the first patent for an electromagnetic telegraph. The telegraph was first used in conjunction with the country's rapidly expanding railway system, as a way of sending signals about traffic down the line. By 1845, the Electric Telegraph Company owned Wheatstone's and Cooke's patent and offered their services to the paying public. Victorian commentators waxed lyrical over the new technology's possibilities. According to Dionysius Lardner, 'of all the physical agents discovered by modern scientific research, the most fertile in its subservience to the arts of life is incontestably electricity, and of all the applications of this subtle agent, that which is transcendently the most admirable in its effects, the most astonishing in its results, and the most important in its influence upon the social relations of mankind, and upon the spread of civilization and the diffusion of knowledge, is the Electric Telegraph'. The telegraph was widely held to have brought about the literal annihilation of time and space. By the end of the 1840s, telegraph networks were rapidly spreading across Britain, continental Europe and the United States. By the 1850s, the first efforts to lay underwater cables linking Britain with the rest of Europe and with Ireland were being made. A new profession was also being born – the telegraph engineer.

By the middle of the 1850s, a scheme was afoot to attempt the most ambitious telegraphic project yet – to lay a telegraph cable under the Atlantic linking the Old World with the New. Pushed forward by the American entrepreneur Cyrus Field, the project brought together engineers, financiers, inventors and natural philosophers of all kinds. Even Faraday was involved, giving his opinion on what design of cable would be best suited for submarine use. The first cable was laid in 1858 and triumphant messages flashed across the Atlantic. Within a month, however, the cable had failed. During the recriminations that followed, William Thomson made a name for himself showing that the problem lay with the telegraph engineers and operators who oversaw the cable's operation. What was needed,

Thomson suggested, was a new kind of telegraph engineer who understood electrical theory. Another unsuccessful effort to lay a transatlantic cable was made in 1865 – this time the cable broke. Success came eventually in 1866, with Isambard Kingdom Brunel's massive steamship, the *Great Eastern*, laying the cable. By the end of the century, underwater telegraph cables spanned the globe, playing a vital role in commerce and diplomacy. Protecting the 'All-Red Route' of telegraph cables that linked together Britain's far-flung imperial possessions was, by the end of the century, a major preoccupation for the Empire's rulers.

Illustration 14: Ships in Plymouth harbour in 1858, being loaded with the cable that would be laid out across the Atlantic to form the first transatlantic telegraph line.

William Thomson's involvement in the Atlantic cable project shows us just how close the links were between electrical scientists like him and telegraph engineers. One of the most important electrical projects of the second half of the century – to define a set of standardised electrical units of measurement – was as consequential for telegraphy as it was for electrical science. In fact, it was this project that Thomson had in mind with his remark about turning electricity into a 'purchaseable tangible object' quoted earlier. Not only Thomson, but James Clerk Maxwell too, played an important role in the project to define electrical units. Cambridge's Cavendish laboratory under Maxwell and his successor Lord Rayleigh became a veritable 'manufactory of ohms'. Defining electrical units was vital for telegraph engineers because they needed accurate measurements in order to be able to find and repair breaks in telegraph cables. Increasingly, too, accurate and precise measurement was the be-all and end-all of physics. National pride was at stake as well. Physicists from different countries battled it out in successive international congresses throughout the 1880s until Britain eventually emerged triumphant with the Cavendish ohm being accepted as the international standard of electrical resistance. Dominance in the field of electrical science and measurement went along with dominance in electrical commerce and industry.

By the 1880s, telegraphy was being eclipsed as the main electrical industry by the burgeoning development of the electric light and power industries. Symbolically, in 1889 the Society of Telegraph Engineers changed its name to become the Institution of Electrical Engineers. Inventors and entrepreneurs like Sebastian di Ferranti in Britain and Thomas Alva Edison in the United States made fortunes through electricity. Across the Atlantic, the so-called 'Battle of the Systems' raged between Edison and his main competitor, George Westinghouse, over the issue of alternating versus direct current for electric power distribution. Edison was a fan of direct current transmission, while Westinghouse favoured alternating current. In a particularly gruesome twist, Edison succeeded in persuading the New York state authorities to adopt alternating current for use in the electric chair – hoping that the public would learn to associate his rival's system with death rather than life. As electrical networks proliferated, a whole panoply of new electrical inventions emerged to tempt consumers into consuming ever more of the electric fluid. By the end of the century, electric lights illuminated city streets and homes, electric trams carried commuters to their work – it was even possible to cook and do the laundry with electric current. Utopian writers prophesied a future in which everything would be done by electricity. This

19th-century electrical world was one of ambitious projects and spectacular new inventions which their inventors hoped would change the world.

It was also a world of flamboyant exhibition. Selling electricity meant making it visible. Following the success of the Great Exhibition of 1851, national and international exhibitions were hugely popular during the second half of the 19th century. Electrical shows were an integral part of this exhibition culture. The fight between Edison and Westinghouse (which Westinghouse won) for the honour of electrifying the Columbian Exposition in Chicago in 1893 is a good illustration of the extent to which *fin de siècle* electrical concerns valued the opportunity such events afforded them of putting their wares before the public. The Electricity Building there was lit by 120,000 electric lights. Visitors could enjoy the novel experience of travelling from building to building around the site on an electric railway. Men like Edison or Ferranti or Westinghouse invested fortunes in putting on bravura electrical displays at such exhibitions. Showmanship like this was an integral part of what it meant to be an electrical inventor at the end of the 19th century. The Serbian-born inventor and showman Nikola Tesla is another good example. Tesla's public lectures were a byword for dramatic display. His high potential, high frequency electrical apparatus could produce a whole array of spectacular lights and amazing sparks and effects of all kinds. The highlight of Tesla's performances was when he placed himself in the circuit of his electricity-generating equipment, holding illuminated lightbulbs in his hands and passing sparks between his fingers. Literally making himself a part of his invention was ideally calculated to demonstrate his own mastery over it. Not at all like Faraday's lecturing style and vision of electricity.

· CHAPTER 27 ·

LAST THOUGHTS

So just what was Michael Faraday's contribution to the electrical century? Traditional views have tended to regard Faraday as towering head and shoulders above his fellow electrical scientists. Certainly, most histories of 19th-century electricity are dominated by Faraday. In many respects, it is easy to see why. He was, after all, a towering figure. His three decades of *Experimental Researches* were, by any standards, a staggering achievement. His impact as a populariser and public performer was profound. He played a central role in making the natural sciences into an important part of Victorian cultural life. He clearly inspired a new generation of experimenters, though, interestingly enough, he never groomed a successor in the same way as Davy had groomed him. It is also undeniable that there is – as some of his contemporaries themselves noted – something seductively romantic about Faraday's story. As we have seen already, it did not take the Royal Institution's President very long to realise that there was 'something romantic and quite affecting in such a conjunction of Poverty and

Passion for Science'. His intellectual rags-to-riches story was just as appealing to a mid- and late-Victorian audience brought up on tales of Smilesian self-help. The story of the dogged outsider battling against the odds and a devious conservative establishment to eventual triumph is clearly resonant today as well.

Faraday was certainly adopted as father figure by the nascent electrical industry during the early years of the 20th century. The Institution of Electrical Engineers' corporate seal bears Faraday's image. The Institution established a Faraday Medal to celebrate major achievements in electrical engineering. The first winner, in 1922, was the Maxwellian theorist Oliver Heaviside. Faraday is still widely hailed as the 'father of electrical engineering'. From this perspective, Faraday's humble electromagnetic rotation device of 1821 became the first electric motor. His electromagnetic induction devices of 1831 became the first electric generators. Faraday's image proved to be a potent one that could be used to help sell electricity to a still suspicious early 20th-century public. His life history, his reputation as an electrical experimenter and his unique position at the Royal Institution all helped make him an attractive founding father for the electrical industry. In the same way, Maxwell's decision during the 1850s to build his own electromagnetic theories around Faraday's work helped solidify Faraday's own standing as a theorist – as well as allowing Maxwell to point to solid empirical foundations for his own speculations. Ironically, many modern commentators point to Faraday's lack of overt theorising as the key to his experimental success. Faraday, they suggest, coming from outside the scientific establishment, had no theoretical axes to grind – he simply saw nature as it really was. This is nonsense, of course. By the time he was an active and successful experimenter, Faraday was firmly ensconced at the heart of the scientific establishment and was proud of his theoretical acumen – whatever others like William Whewell thought of it before Maxwell came along.

These kinds of images of Faraday as a unique and isolated figure in the history of 19th-century electrical science and culture end up obscuring, rather than illuminating, his real place in a highly complex scientific culture. By focusing on Faraday, we can very easily end up missing the role and importance of other figures and other ways of going on. Faraday himself had a tendency to undervalue the contributions of some other contemporaries – as at least one of his biographers noted. Silvanus P. Thompson suggested that Faraday had never given William Sturgeon full credit for his invention of the electromagnet – an innovation that Thompson placed on a par with Faraday's own discovery of electromagnetic induction. Given the evident bad blood between Faraday and Sturgeon – and their mutually antipathetic views of what experiment and electricity were all about – maybe this is hardly surprising. But Sturgeon is not the only figure to disappear in Faraday's shadow. Even someone like James Clerk Maxwell ends up being relegated to the margins of scientific history by the single-minded focus on Faraday's place. Paradoxically, an undue focus on Faraday ends up diminishing his own achievement as well. If we forget the alternatives that there were to Faraday's interpretations of nature and his way of doing things, then we forget the hard work he had to do to win the arguments.

Faraday himself would have found some aspects, at least, of his modern reputation hard to fathom – and hard to stomach too. He would, in all likelihood, have been insulted, rather than flattered, to be described as the inventor of the electric motor or generator. Like many of his contemporary men of science – in more gentlemanly circles, at least – Faraday drew a strict distinction between discovery and invention. Discovery was what philosophers did – invention was a far lesser and far more sordid affair. He would have agreed with his friend William Robert Grove's view that those who sold inventions for monetary gain had no right 'to look for fame' as well. There is no doubt that philosophical fame rather than material fortune is what Faraday was after. The suggestion that he was devoid of theory is another that Faraday himself would have found a little odd. While he might not go all the way with his master Humphry Davy's German-inspired views about the transcendent Unity of Nature, he was certainly aware of them and sympathetic to their implications. This, after all, is why he was interested in the relationship between electricity and magnetism in the first place. He would probably have agreed with the view that he was a self-made man who had dragged himself up by his bootstraps, but would have been less sure that he offered anyone else a good role model.

In the end, what the focus on Faraday (or any other individual) obscures is the fact that science is a collective rather than an individual enterprise. Victorian Britain had a complex and vibrant scientific culture. It was full of radicals and ne'er-do-wells who wanted to use science – and electricity in particular – to overthrow civil society. It also sported more sober sorts who thought that science could be a bulwark against social disintegration. Victorian science was full of people on the make, trying to survive in a competitive world by re-making science in their own image. Science could be a source of fame and social status; it could also be a way of cashing in. Disagreements raged – about the relationship between discovery and invention; about the relationship between science and the state, or science and religion; about what kind of person the man of science (and it was still usually a man) should be, and what kinds of institutions he should work in. Following Faraday around – as this book tries to do – provides one useful way of looking at this complex culture. But it should not blind us to the fact that, however important he was, Faraday was only one player in the game. It took far more than one man's genius to forge the electrical century.

Further Reading

Prologue

Two of the best 19th-century biographies of Faraday are Henry Bence Jones, *The Life and Letters of Faraday*, 2 volumes (London, 1869) and John Tyndall, *Faraday as a Discoverer* (London, 1868). The standard scientific biography is L. Pearce Williams, *Michael Faraday* (London: Chapman and Hall, 1965). The most recent accessible biography is James Hamilton, *Faraday: The Life* (London: HarperCollins, 2002). A great deal of information about Faraday, his contemporaries and 19th-century scientific culture can be found in Frank James (ed.), *The Correspondence of Michael Faraday*, 4 volumes (London: Institution of Electrical Engineers, 1991–9). For some useful background on 18th-century electricity and culture, see Patricia Fara, *An Entertainment for Angels: Electricity in the Enlightenment* (Cambridge: Icon Books, 2002).

Growing Up in Scientific London

Early 19th-century scientific London is surveyed in Iwan Rhys Morus, James Secord and Simon Schaffer, 'Scientific London', in Celina Fox (ed.), *London – World City* (New Haven CT and London: Yale University Press, 1992). Early 19th-century scientific societies in the metropolis are examined in Ian Inkster, 'Science and Society in the Metropolis: A Preliminary Examination of the Social and Institutional Context of the Askesian Society of London, 1796–1807', in *Annals of Science*, 1977, **34**, pp. 1–32. For more on the London lecturing circuit, see J.N. Hays, 'The London Lecturing Empire, 1800–50', in Ian Inkster and Jack Morrell (eds), *Metropolis and Province: Science in British Culture, 1780–1850* (London: Hutchinson, 1983) and J.N. Hays, 'Science in the City: the London Institution, 1819–40', in *British Journal for the History of Science*, 1974, **7**, pp. 146–62. For Faraday and the City Philosophical Society, see Frank James, 'Michael Faraday, the City Philosophical Society and the Society of Arts', in *RSA Journal*, 1992, **41**, pp. 192–9.

The Philosopher's Apprentice

For some background on Davy, see June Fullmer, *Young Humphry Davy: The Making of an Experimental Chemist* (Philadelphia PA: American Philosophical

Society, 2000) and Jan Golinski, *Science as Public Culture: Chemistry and Enlightenment in Britain, 1760–1820* (Cambridge: Cambridge University Press, 1992). For Faraday and Davy's troubled relationship, see David Knight, 'Davy and Faraday: Fathers and Sons', in David Gooding and Frank James (eds), *Faraday Rediscovered: Essays on the Life and Work of Michael Faraday, 1791–1867* (London: Macmillan, 1985). For Faraday's rotation experiments, see David Gooding, 'In Nature's School: Faraday as an Experimentalist', in Gooding and James (eds), *Faraday Rediscovered*. Davy's Presidency of the Royal Society is discussed in David Miller, 'Between Hostile Camps: Sir Humphry Davy's Presidency of the Royal Society of London, 1820–1827', in *British Journal for the History of Science*, 1983, **16**, pp. 1–47. Faraday's rise through the echelons of the Royal Institution is examined in Sophie Forgan, 'Faraday – From Servant to Savant', in Gooding and James (eds), *Faraday Rediscovered*.

Radical Electricity

For a view of radical London, see Iain McCalman, *Radical Underworld: Prophets, Revolutionaries and Pornographers in London, 1795–1840* (Cambridge: Cambridge University Press, 1988). For radical electricity, see Iwan Rhys Morus, *Frankenstein's Children: Electricity, Exhibition and Experiment in early Nineteenth-century London* (Princeton NJ: Princeton University Press, 1998). For Andrew Crosse and his insects, see James Secord, 'Extraordinary Experiment: Electricity and the Creation of Life in Victorian England', in David Gooding, Trevor Pinch and Simon Schaffer (eds), *The Uses of Experiment* (Cambridge: Cambridge University Press, 1989). For Halse and medical electricity, see Iwan Rhys Morus, 'A Grand and Universal Panacea: Death, Resurrection and the Electric Chair', in Iwan Rhys Morus (ed.), *Bodies/Machines* (Oxford: Berg Publications, 2003).

Royal Institution Science

For the Royal Institution, see Morris Berman, *Social Change and Scientific Organization: The Royal Institution, 1799–1844* (London: Heinemann, 1978). Faraday's lecturing style is discussed in Iwan Rhys Morus, 'Different Experimental Lives: Michael Faraday and William Sturgeon', in *History of Science*, 1992, **30**, pp. 1–28. His relationship with Benjamin Abbott is the topic of Frank James, 'The Tales of Benjamin Abbott: A Source for the Early Life of Michael Faraday', in *British Journal for the History of Science*, 1992, **25**, pp. 229–40. On mesmerism, table turning and Faraday's response, see Alison Winter, *Mesmerized: Powers of Mind in Victorian Britain* (Chicago IL: University of Chicago Press, 1998).

Cultures of Display

For the London entertainment scene, see Richard Altick, *The Shows of London* (Cambridge MA: Harvard University Press, 1978). For scientific shows, see Iwan Rhys Morus, 'Manufacturing Nature: Science, Technology and Victorian Consumer Culture', in *British Journal for the History of Science*, 1996, **29**, pp. 403–34. The London Electrical Society is examined in Iwan Rhys Morus, 'Currents from the Underworld: Electricity and the Technology of Display in early Victorian England', in *Isis*, 1993, **84**, pp. 50–69. Later electrical exhibitions are the topic of Carolyn Marvin, *When Old Technologies were New* (Oxford: Oxford University Press, 1988) and Ken Beauchamp, *Exhibiting Electricity* (London: Institution of Electrical Engineers, 1997).

The Great Experimenter

For some background on the history of experimentation, see John Henry, *Knowledge is Power* (Cambridge: Icon Books, 2002). On Faraday's induction experiments, see Sydney Ross, 'The Search for Electromagnetic Induction, 1820–1831', in *Notes and Records of the Royal Society*, 1965, **20**, pp. 184–219. For the Italian controversy, see Brian Gee, 'Faraday's Plight and the Origins of the Magneto-electric Spark', in *Nuncius*, 1990, **5**, pp. 43–68. On

Faraday's electrochemistry and Sturgeon's riposte, see Iwan Rhys Morus, 'The Sociology of Sparks: An Episode in the History and Meaning of Electricity', in *Social Studies of Science*, 1988, **18**, pp. 387–417. For the magneto-optic effect, see David Gooding, 'He who Proves Discovers: John Herschel, William Pepys and the Faraday Effect', in *Notes and Records of the Royal Society*, 1985, **39**, pp. 229–44. For Faraday's theoretical views and their relationship to his experiments, see David Gooding, 'Magnetic Curves and the Magnetic Field: Experimentation and Representation in the History of a Theory', in Gooding, Pinch and Schaffer (eds), *The Uses of Experiment*.

The Electrical Century

For Faraday's Sandemanianism, see Geoffrey Cantor, *Michael Faraday: Scientist and Sandemanian* (London: Macmillan, 1991). For Faraday and Maxwell's physics, see Peter Harman, *The Natural Philosophy of James Clerk Maxwell* (Cambridge: Cambridge University Press, 1998). For Maxwell's Maxwellian successors, see Bruce Hunt, *The Maxwellians* (Ithaca NY: Cornell University Press, 1991). For telegraphy, see Iwan Rhys Morus, 'The Electric Ariel: Telegraphy and Commercial Culture in Early Victorian England', in *Victorian Studies*, 1996, **39**, pp. 403–34 and Bruce Hunt, 'The Ohm is where the Art is: British Telegraph Engineers and the Development of Electrical Standards', in Albert van Helden and Thomas Hankins (eds), *Instruments*, special issue, *Osiris*, 1994, 9, pp. 48–63. On the rise of the electrical industry, see Thomas Hughes, *Networks of Power: Electrification in Western Society* (Baltimore MD: Johns Hopkins University Press, 1983) and David Nye, *Electrifying America: Social Meanings of a New Technology* (Cambridge MA: MIT Press, 1990).

Sources for Illustrations

1. Henry Bence Jones, *The Life and Letters of Faraday*, vol. 1.
2. Wilkinson, *Theatrum Illustrata*, 1825.
3. British Museum, Department of Prints and Drawings.
4. Michael Faraday, *Experimental Researches in Electricity*, vol. 2.
5. Figuier, *Les Merveilles de la Science*, vol. 1.
6. *Annals of Electricity*, vol. 2.
8. *Illustrated London News*, 1856.
10. *Year-book of Facts in Science and Arts*, vol. 3.
11. Henry Bence Jones, *The Life and Letters of Faraday*, vol. 2.
12. Michael Faraday, *Experimental Researches in Electricity*, vol. 1.
13. *Philosophical Magazine*, vol. 21.
14. *Illustrated London News*, 1858.